Lecture Notes in Economics and Mathematical Systems

490

Springer
Berlin
Heidelberg
New York
Barcelona
Hong Kong
London
Milan
Paris
Singapore
Tokyo

Stefan Minner

Strategic Safety Stocks in Supply Chains

Springer

Author

Dr. Stefan Minner
Faculty of Economics and Management
Otto-von-Guericke-University Magdeburg
Universitätsplatz 2
39106 Magdeburg, Germany

Cataloging-in-Publication data applied for

Die Deutsche Bibliothek - CIP-Einheitsaufnahme

Minner, Stefan:
Strategic safety stocks in supply chains / Stefan Minner. - Berlin ;
Heidelberg ; New York ; Barcelona ; Hong Kong ; London ; Milan ; Paris
; Singapore ; Tokyo : Springer, 2000
 (Lecture notes in economics and mathematical systems ; 490)
 ISBN 3-540-67871-9

ISSN 0075-8442
ISBN 3-540-67871-9 Springer-Verlag Berlin Heidelberg New York

Springer-Verlag Berlin Heidelberg New York
a member of BertelsmannSpringer Science+Business Media GmbH

© Springer-Verlag Berlin Heidelberg 2000
Printed in Germany

Typesetting: Camera ready by author
Printed on acid-free paper   SPIN: 10778516   42/3142/du   543210

Preface

The management of supply chains has become one of the key issues in logistics management over the recent decade. On the one hand, this development is caused by the extraordinary progress in information technologies and communication systems that have developed over the last years. On the other hand, it is also due to the enormous potentials in improving logistics efficiency that can be exploited if an intelligent coordination of material flows and stock control decisions in large supply chains is practiced. This task of effective coordination of materials management decisions is an extremely challenging one, regarding the fact that usually supply chains in practice turn out to be complex supply webs and logistics processes within these networks regularly are influenced by various uncertainties.

A scientific field offering valuable tools to support decision making in such a supply chain context is the theory of multi-stage inventory control. This is because material coordination in supply webs has to be carried out from an integral perspective of all material procurement and distribution decisions that refer to the different stock points over all stages of a network. Hereby, logistics performance is affected by the internal cost at which the whole system is operating and by the service that it can provide to external customers. A major problem in this context that until now has not been sufficiently solved despite of all advances in multi-stage inventory theory is the following: How should the size and the allocation of safety stocks in complex supply networks with stochastic customer demands be determined in order to guarantee specified consumer service levels at minimum inventory holding costs?

This book presents a break-through in research on this issue by extending theoretical analysis and optimisation techniques for multi-stage safety stock planning to such a level that supply webs of arbitrary size and structural complexity, even including cyclical network parts, can be handled with limited computational effort. This holds for situations where safety stocks are determined from a strategic perspective which takes into consideration that in supply networks usually some operational flexibility exists which can be used as an additional means of short-run protection against uncertainties which may cause material shortages.

Overall, the contribution of this book is two-fold. It presents an excellent scientific work offering a major progress in the field of advanced stochastic inventory theory, and it delivers effective algorithms for cost-minimising safety stock determination in highly complex supply networks that can be applied in practice. So it can provide managers with a promising scientific tool for solving one of the very challenging problems in modern supply chain planning.

Magdeburg, May 2000   *Karl Inderfurth*

Acknowledgements

Many people have contributed valuable support during the process of writing this thesis at the Faculty of Economics and Management at the Otto-von-Guericke-University of Magdeburg. First, I would like to express my gratitude towards my supervisor Prof. Dr. Karl Inderfurth for motivating the research on safety stock problems and many helpful suggestions and discussions. I owe more than only thanks for their assessment on the thesis to Prof. Dr. Alfred Luhmer and Prof. Dr. Ton de Kok. During my business administration studies at the University of Bielefeld, Prof. Dr. Alfred Luhmer mainly inspired my interest in Operations Research which was intensified by Prof. Dr. K.-P. Kistner. During a three month research visit in Eindhoven, I enjoyed Ton's enthusiasm for multi-echelon inventory models and learned a lot of from the joint research work with him and Erik Diks. For many research discussions, mental support, and a stimulating working climate I am indebted to my former colleagues at the University of Magdeburg, Gerald Heisig, Dr. Thomas Jensen, and Dr. Dirk Meier-Barthold. Rainer Kleber provided a lot of technical support during the final writing phase and Ian M. Langella made several suggestions for improving the English. Last but not least I want to thank my parents for supporting my education and Steffi for her sympathy for the time I spent on finishing this thesis.

Magdeburg, May 2000   *Stefan Minner*

Contents

1. Introduction

Increasing customer satisfaction requirements concerning products and service, more differentiated customer desires, and international competition, determine new opportunities and requirements for the management of logistics processes. The diversity of customer needs with it's product variety leads enterprises to develop new strategies like mass customization in order to deal with the organizational complexity and cost increases. Within the area of supply chain management, concepts for the coordination of such systems are developed, discussed, compared, and implemented. The coordination concerns different internal processes as well as the integration of external suppliers and customers and is supported by recent developments in information and communication technology. Within the efficient consumer response (ECR) philosophy, the coordination of logistics processes is mainly directed to customer requests by collecting point-of-sale (POS) information (e.g. with scanner terminals) and by immediate transmission to all involved supply chain processes and partners using electronic data interchange (EDI) technology or even by utilizing the internet. Practical applications derived from the ECR-philosophy are the continuous replenishment process (CRP) and vendor managed inventories (VMI).[1]

Nevertheless, these popular materials coordination approaches lie within well known concepts of materials and inventory management for multi-echelon systems. Available concepts are generally divided into program oriented approaches like Material Requirements Planning (MRP) and Distribution Requirements Planning (DRP) on the one hand and stochastic inventory control (SIC) on the other hand. The second category is further divided into systems being locally controlled by installation stock policies and those controlled under central information with echelon stock policies. The ECR-philosophy provides systemwide information to all involved supply chain agents for their replenishment decisions.

Within all coordination concepts, the production and replenishment planning is affected by uncertainty introduced through customer demand variability and production and supply process unreliability. These disturbances result from time and quantity deviations of realized from planned data. Deviations in time include delays in customer interarrival times and replenishment lead

---

[1] FISHER, M.L. [1997].

times. Quantity deviations show up within customer order quantities and the quality of supply deliveries. In order to protect customer service performance on the external interface of the system and, in addition, to decouple operations from unforeseen deviations and planning adjustments, buffers are introduced. Available mechanisms are to hold stocks in addition to the average requirements (planned safety stocks), to release orders and to initiate production processes earlier than necessary under average supply and processing conditions (planned safety lead times), to reserve additional processing capacities for the introduction of flexibility that enables the processes to deal with extraordinary large deviations without affecting the performance outcome, and to overestimate the average forecasted requirements.[2]

Some of the pitfalls observed in practical supply chains are insufficient information systems and coordination, unsatisfactory incorporation of buffers, especially the fact that buffer sizes are often chosen independently from the underlying variability, and incomplete incorporation of all relevant supply chain processes.[3] The improvement of these shortcomings can offer substantial cost reductions and service improvement potentials. Therefore, large research efforts have been directed to materials coordination problems. Bibliographies of inventory models are provided by LEE, NAHMIAS and CHIKAN for a single inventory point[4] and by VAN HOUTUM, INDERFURTH, ZIJM, and DIKS, DE KOK, LAGODIMOS, and FEDERGRUEN, and AXSÄTER for multi-echelon systems with interacting inventory points.[5]

The applied research methodology in the field of multi-echelon systems inventory control can roughly be divided into three categories: (1) rules of thumb, (2) simulation analysis, and (3) analytic models. The first category contains more or less justified rules for buffer mechanism choice, allocation, and the respective sizing of buffers. For buffer mechanism choice, one rule that has been justified later by simulation and analytic work is to buffer against time uncertainties by time buffers (e.g. safety times) and to deal with quantity uncertainties by planned safety stocks. Rules of thumb concerning buffer allocation propose to hold buffer inventories only at the final level or alternatively at the first and the final level, i.e. at the interface stockpoints of the system to its environment. Simulation models and studies serve to model the complex dependencies in a multi-echelon system and to gain insights of different buffering alternatives on system performance measures.[6] The advantage of this approach is that it is in general not limited by too restrictive assumptions whereas it is rather time consuming for product structures of practical size. Therefore, the conclusion appears that fast heuristic models

---

[2] WIJNGAARD, J., WORTMANN, J.C. [1985].

[3] LEE, H.L., BILLINGTON, C. [1992].

[4] LEE, H.L., NAHMIAS, S. [1993], CHIKAN, A. [1990].

[5] VAN HOUTUM, G.J., INDERFURTH, K., ZIJM, W.H.M [1996], DIKS, E.B., DE KOK, A.G., LAGODIMOS, A.G. [1996], FEDERGRUEN, A. [1993], AXSÄTER, S. [1993].

[6] See e.g. CHAKRAVARTY, A.K., SHTUB, A. [1986], MOLINDER, A. [1997].

are necessary. The third stream of analysis uses analytic models in order to describe the dependencies and to provide decision support for the solution of buffer choice, allocation, and sizing problems. The price being paid for obtaining exact analytic results is often the incorporation of limiting, non-realistic assumptions, e.g. two-stage supply chains, negligible (zero) lead times, or identical inventory points. Therefore, these insights and the provided support is limited to special practical applications and settings and will, in general, require completely different models for other applications under the same materials coordination background.

An approach that aims to provide decision support for a large variety of applications has to fulfill some basic requirements.[7] The approach should be able to deal with generalized product structure networks instead of being limited to special topologies. A joint incorporation of uncertainties with respect to supply, processes, and demand is required in spite of assuming a single source of variability. In addition, the approach still should be simple for organizational implementation, understandable, and tractable for analysis and computation. Finally, the application within a capacitated production system or at least the extendability to this environment is necessary. It is obvious that exact analytic inventory models will not be able to fulfill all these requirements simultaneously. Nevertheless, suitable approaches that do not explicitly provide assistance in all mentioned situations should at least be extendable to these settings without destroying the general idea behind the model. A unified framework instead of the development of special models for every single type of problem represents a necessary condition for implementation in standard software packages.

The target of this thesis is to present and extend a multi-echelon inventory coordination approach that provides (heuristic) decision support for adequate safety stocks in practical applications of general network supply chains. A hierarchical planning concept[8] is followed with respect to two aspects. First, the problem is decomposed into a higher level safety stock coverage allocation problem that is solved under the incorporation of lower level safety stock sizing problems for each inventory point in the system. Then, simple single-echelon models with the relevant problem features are coordinated by a multi-echelon model.[9] Second, implemented buffers represent strategic safety stocks meant to provide slack for normal variations in problem data. Extraordinary large deviations are assumed to be offset by operative (emergency) activities. The information flow required to fulfill the model assumptions is similar to the ECR-philosophy, i.e. point-of-sale data is immediately transmitted to all supply chain processes that are involved in the replenishment process for the respective product.

---

[7] LEE, H.L., BILLINGTON, C. [1993].

[8] SCHNEEWEISS, C. [1999].

[9] This idea is similar to the suggestion of LEE, H.L., BILLINGTON, C. [1993].

The multi-echelon safety stock planning problem consists of several economic problem components. From a production theory point of view, the required customer service can be produced by several safety stock allocations. If processing adds value to the products, the corresponding holding cost of safety stocks is cheaper at previous stages compared to downstream placement of safety stocks. Safety stocks held at some inventory point in order to provide the required customer service can be substituted by increasing upstream safety stocks, i.e. buffers being immediately available for customer satisfaction are substituted by better replenishment service. The economic goal is to find the minimal cost combination that satisfies the service requirements. The second aspect concerns the exploitation of economies of scale. If several single (independent) random numbers are aggregated, the variability of the aggregate only increases with the square root of the number of random variables. Therefore, buffer aggregation profits from these economies of scale, whereas buffer allocation requires other advantages to compensate the loss of economies of scale, namely lower holding cost and the portfolio effect of risk pooling. If a stockpoint supplies several products, these products represent the supply portfolio and the joint variability of this portfolio is smaller than the sum of the single variabilities (except for the case of perfect positive correlation).

The following chapters present the fundamental principles of safety stock planning and materials coordination and the developed safety stock planning approach for general supply chains. In Chapter 2, single-echelon inventory models are presented with respect to the main application components and respective modeling alternatives. For given service level constraints, models to determine the required safety stock levels with respect to uncertain demand and processing times and under lot-for-lot as well as under batch replenishment are presented. Multi-echelon system types, together with a discussion of possible application backgrounds, are introduced in the third chapter. Additionally, different concepts for materials coordination in supply chains are presented and the role of safety stocks within the respective approach is discussed. Chapter 4 represents the main part of the thesis where alternative approaches towards the safety stock planning problem are presented for serial, divergent, and convergent systems, and then extended to general network supply chains. The section on serial systems starts with the model of CLARK, SCARF[10] where safety stocks are assigned to cover against all demand variations. The respective materials flow and an algorithm for the determination of allocation and size of safety stocks is reviewed. In contrast, the model of SIMPSON[11] that represents the fundamental basis of the extensions developed within this thesis, assumes that safety stocks only hedge from normal variability. The materials flow, the resulting optimization problem and its properties are presented before several Dynamic Programming for-

[10] CLARK, A.J., SCARF, H. [1960].
[11] SIMPSON, K.F. [1958].

mulations and local search based heuristics are outlined to solve the safety stock planning problem. The two approaches are compared and an idea for a synthesis that combines the advantages of both approaches is provided. Required extensions of the base model in order to provide sufficient decision support in more general serial supply chains with different service measures, stochastic processing times, and batch replenishment are presented afterwards. Additional problems that appear if an inventory point faces several successor requirements in a divergent system or several predecessor supply points in a convergent system, resulting adjustments within the optimization problem, and solution procedures are outlined in Sections 4.2 and 4.3. If both aspects of multiple predecessors and successors are present simultaneously, the required extensions to the optimization problem are given in 4.4. A general Dynamic Programming approach is introduced before several heuristics that are based on popular meta search strategies are fitted to the safety stock planning problem. Afterwards, the algorithms are compared in a short numerical experiment. Finally, the effect of cycles resulting from external product returns with remanufacturing operations or introduced by internal by-product reuse options, is investigated and an idea for the incorporation into the logic of the SIMPSON model is outlined.

2. Single-Echelon Inventory Models

This chapter consists of four parts. In the first part, different systems and motives of inventory control are presented. Physical stocks in an inventory system are categorized and the sources of uncertainty are outlined. In the second part, a single-echelon, single-product inventory system is described by its main influencing factors. In parallel, approaches for modeling different system characteristics are presented. Quantitative safety stock planning procedures that are required for the different models are summarized in the third part. Finally, a different approach in stochastic inventory theory that utilizes the feedback concept instead of applying feedforward control of performance measures is mentioned.

The presented material focuses on the situation of a single product processed by a single processor, e.g. a purchasing or production process. The text book knowledge[1] summarized in this chapter serves as a building block for the materials coordination in multi-echelon systems.

2.1 Terminology and Classification

2.1.1 Classification of Inventory Control Systems

Though the terminology "inventory control system" is widely used, it implies singularity, whereas in reality there are several models, necessitated by different aspects and concerns. Three kinds of inventory control systems are distinguished between:[2]

- Inventory system
  The most simple type is a pure inventory system where inventory control decisions for different products are made independently. This approach is applied for the replenishment of individual items. Competition among multiple products for scarce capacity and the interaction that results from consecutive stages of processing are not taken into account.

---

[1] SILVER, E.A., PYKE, D.F., PETERSON, R. [1998], HAX, A.C., CANDEA, D. [1984].

[2] HAX, A.C., CANDEA, D. [1984], p. 127-128.

- Production/inventory system
  In production/inventory systems, at least one of the aspects being omitted in the analysis of pure inventory systems is considered, that is joint utilization of capacity or multi-echelon processing dependencies.
- Distribution inventory system
  While the previous two types of inventory systems are applied to manufacturing and assembly oriented problems, a distribution system covers the problem of allocating products among several stocking points. Depending on the kind of replenishment decision making and the information available at the stocking points, push and pull control systems are distinguished between. In a push control system, central information about the inventory status of the stocking points is available and used as a basis for a central allocation decision. In a pull system, every stocking point decides upon releases of an order at the supplying installation on the basis of its local information.

2.1.2 Motives of Inventory Control and Stock Classification

Stocks in inventory control systems are induced by decisions which themselves are driven by several motives. The motives can roughly be divided into transaction, safety, and speculation motives. The transaction motive is a result from the fact that ordering and production decisions are carried out at certain points of time instead of being performed continuously. The safety motive appears when some of the required data, e.g. lead times, demand, production yield, is unknown when decisions have to be made. Nevertheless, system performance and smooth operation should not be significantly disturbed by deviations from planned data and therefore, buffers are introduced. The speculation motive in general refers to a special kind of uncertainty, mostly in prices. Due to the anticipation of increasing prices for purchased products, requirements are ordered in advance.

Based on these motives, inventories are classified into five categories.[3]

- Cycle stocks
  The transaction motive causes ordering and production in batches. The cycle stock induced by batching alternates between an upper level just after the arrival of a batch and a lower level just before the arrival of the next batch. The reasons that induce cycle stocks are in most situations economies of scale and technological restrictions. The presence of a major setup cost for an order or production release or a concave cost function favor a batching of material requirements. The same effect is achieved by quantity discounts offered by the supplier. In other cases that are not motivated by economic reasons, a just in time lot-for-lot replenishment is

---

[3] SILVER, E.A., PYKE, D.F., PETERSON, R. [1998] p. 30-32, HAX, A.C., CANDEA, D. [1984], p. 125-127.

restricted by technological constraints. For instance, some minimum production quantity is necessary to enable the processing.

- Pipeline stocks
  The fact that order processing times, production, and transportation rates are finite causes process inventory. This category includes all material that is in process as well as material that is transported and in transit to another processing unit. Processed material of a batch that is not finished and is waiting until the batch is completed and can be shipped to its destination belongs to the pipeline stock instead of being assigned to cycle stocks.

- Safety stock
  The safety stock is defined as the expected inventory just before the next order arrives. This buffer stock is induced by the uncertainty of some problem components like demand, processing time, yield. For planning purposes, estimations for these uncertainties are made but forecast errors can extremely dampen system performance. Therefore, safety stocks are implemented in order to protect system performance from forecasting errors. A broader interpretation of safety stocks is given by HAX, CANDEA. In their interpretation, safety stocks are held for situations when ordered deviates from delivered, both in time and in quantity.[4] In extension to the forecasting error interpretation, this also includes delivery delays caused by capacity constraints.[5]

- Speculative stocks
  Expected price increases may result in earlier replenishments than would be experienced under constant purchase prices. Additionally, profits gained by selling purchased products at a higher price induce speculative stocks. In contrast to the transaction motive, speculative stocks result from expectations rather than from quantity effects.

- Anticipation stocks
  Seasonal phenomena, rather than expectations, generate anticipation stocks. A time varying demand pattern for products requires for balancing of overtime and inventory holding cost in order to cope with demand peaks. The same production smoothing argument applies under seasonal demand and convex production cost functions. A different problem of seasonality on the supply side is found for agricultural products, which may only be available at some periods of time or only at higher prices for the remainder of the year. The anticipation of these seasonal effect results in anticipation stocks. Under a generalized point of view, speculative stocks can also be regarded as stock induced by anticipation rather than by expectation of price increases.

The planning of stock norms is often connected to the above classification. Models are developed for each category independently, that is lot-sizing,

---

[4] HAX, A.C., CANDEA, D. [1984], p. 126. This definition is coupled to the classification of uncertainty to be presented in the next subsection.

[5] GRAVES, S.C. [1988].

safety stock, and production smoothing models. Nevertheless, it is difficult to determine to which of the categories a certain item belongs. This problem raises from the fact that stocks can originate from more than a single inventory control motive and that a certain degree of substitution is present. An inventory created by a batching decision can at the same time serve as a buffer because the excess cycle stock implies a certain amount of safety potential to cover against forecasting errors.

SILVER, PYKE, PETERSON discuss another category of decoupling stock.[6] They define decoupling stock as the inventory that is necessary to permit the separation of decision making in multi-echelon system. On the other hand, this stock can be regarded as a buffer against internal demand uncertainty and can be grouped into the safety stock category. In case of lot-size integrality induced stocks, this inventory belongs to transaction motivated cycle stocks.

For accounting purposes, turnover ratios are defined as the fraction of sales per year and inventory in one of the categories raw material, work-in-process, and finished goods and serve as performance indicators.

- Raw material
  Inventory that has been purchased and delivered from external suppliers and that is waiting for the first processing or assembly is grouped into this category. It includes cycle stocks from supply batches, safety stocks that are held to protect against supply uncertainty, anticipation stocks, and speculative stocks.
- Work-in-process
  Work-in-process consists of manufactured parts, assemblies, and subassemblies. These unfinished products have been processed at least by one operation. The difference between this and the definition of pipeline stock results from decoupling (safety) stocks not being in transit or being processed.
- Finished goods
  Finished goods include all items stored after the final processing step that are waiting for customer market requirements or delivery operations. These stocks either reached their destination point in the system as cycle stock due to batches at the final operation, or are held as safety stocks, or imply seasonal inventory due to production smoothing.

The use of this classification for inventory control purposes is nevertheless rather limited.

2.1.3 Uncertainty and Planning Techniques

The scope of this section is to classify uncertainty sources in inventory systems and to describe how uncertainty is addressed in different inventory control models. According to WHYBARK, WILLIAMS[7], uncertainty is classified with

---

[6] SILVER, E.A., PYKE, D.F., PETERSON, R. [1998], p. 31.
[7] WHYBARK, D.C., WILLIAMS, J.G. [1976], p. 598-600.

respect to source and effect. Uncertainty can result either from demand or from supply processes. In a single-echelon system, these sources are external customers and suppliers whereas in a multi-echelon system, the corresponding sources are succeeding stockpoints which induce internal requirements on the demand side and preceding stockpoints on the supply side.

Within demand and supply sources, uncertainty can be further divided into timing and quantity effects. Demand timing uncertainty occurs when customers or internal material recipients release their orders earlier or later than predicted. When the size of their material requests deviates from the forecasts, demand quantity uncertainty occurs. Additionally, both effects can occur simultaneously. Within the modeling approach of periodic review inventory control systems, where a probability distribution is determined for single period demand, both effects are mixed together. The stochastic process modeling of continuous review systems with an interarrival process for orders and a separate probability distribution of each order size enables to model both effects explicitly.

Supply timing uncertainty results from deviations in the supplier delivery time or from the fact that internal orders have been rescheduled by reassigning priorities for orders at bottleneck work stations. Therefore, planned order receipts occur earlier or later in time. Yield losses, partly insufficient quality of delivered products, limited materials availability, or scarce capacity at the external or internal supplier results in quantity deviations and therefore, in supply quantity uncertainty. Especially the situation of scarce supplier capacity may result in the delivery of subbatches and therefore, in timing and quantity uncertainty simultaneously. As a consequence, the possibility and the usefulness of a separate modeling of time and quantity deviations is often limited.

The presence of uncertainty requires an appropriate incorporation into the applied planning techniques. Available planning concepts in inventory control contain Stochastic Dynamic Programming procedures and rolling horizon framework based approaches. Stochastic Dynamic Programming requires perfect information on possible future states where only the transition between the states is unknown. The result of this solution technique is a preliminary decision for each future state which tells the decision maker what to do if a certain state is reached. The disadvantage is that a large number of future states can make the planning effort intractable as well as the fact that new possible states may appear in the future, connected with a replanning necessity. Nevertheless, most of the well known stochastic inventory control rules are of this type, namely they advise the decision maker when (in which of the future states) to order how much (conditional decision). Opposed to Stochastic Dynamic Programming, the rolling horizon framework is a mainly deterministic planning concept and therefore requires forecasts about the future development. Based on this data, decisions are planned each period where the decision for the next period is implemented and planned actions for future periods are preliminary. This procedure is repeated every period with an extended planning horizon of one period and adjusted data with respect to forecast updates. According to the influence of uncertainty, a replanning and adjustment of the preliminary decisions takes place. The widely used MRP framework presented in Chapter 3.2 for coordinating material flows in a multi-echelon system utilizes a rolling horizon framework in order to cope with uncertainty.

2.2 Single-Product Inventory Modeling

Following the definition of HAX, CANDEA, an inventory system is given by the coordinated set of rules and procedures that allow for routine decisions on when and how much to order of each item that is needed in the manufacturing or procurement process to fill customer demand.[8] In this section, the modeling ingredients for a single-echelon, single-product production/inventory system are presented. Such a system is characterized by a stockpoint where the items of the corresponding product are stored. The system is described by the state of the stockpoint and is influenced by an input and an output process. The output process is influenced by customer demands or, within a multi-echelon context, by material requirements of succeeding stockpoints. The demand process is modeled in Section 2.2.1.

Inventory of products is replenished from an external supplier or from preceding processing units in a supply chain context. This input process is characterized by the replenishment lead time, that is the time span between ordering of material and receipt of the goods, and by the input coefficient. The input coefficient represents the number of products that are needed from the supplier or the preceding stockpoint in order to produce one item of the product assigned to the stockpoint. The different modeling approaches for the lead time and the input coefficient are discussed in Sections 2.2.2 and 2.2.3 respectively.

The input and the output process are coordinated by an inventory control rule. Basic inventory control rules are presented in Section 2.2.4. These simple rules have in common that they contain two policy parameters. The coordination results are measured by the necessary inventory at the stockpoint and by the customer demand satisfaction. Criteria for performance evaluation are discussed in Section 2.2.5.

2.2.1 Demand

A first characteristic for modeling the demand process concerns the customer behavior according to stock insufficiencies to fulfill all demands.

---

[8] HAX, A.C., CANDEA, D. [1984], p. 129.

- Lost customer
  The demand is lost for the stockpoint and in addition, the customer will not reconsider the selling stockpoint for his future requirements. The modeling of this customer reaction is difficult within standard inventory control models since it requires for an adjustment of the demand model.
- Lost sales
  In the lost sales case, the unsatisfied demand is lost, but there is no effect on future customer behavior.
- Backordering
  In the backorder case, the customers are willing to wait until the missing items become available to the stockpoint. Then, the backordered items can be delivered on a FIFO (first in first out) or on a LIFO (last in first out) basis.[9] Under the FIFO rule, the oldest backorder is satisfied first, whereas under LIFO, the last backorder is satisfied first.

Most inventory models consider either the backorder or the lost sales case. In reality, customer behavior will partly result in lost sales and partly in backorders.

The second component in modeling demand concerns the demand size. In the literature, two streams can be found in general, approaches for fast and slow moving items. The demand process for slow moving items is often separated into an interarrival process of customers and a distribution of the order size. For fast moving items, most approaches assume a demand distribution for the cumulative amount of requested products within in a single period. According to observed demands in previous periods, a theoretical demand distribution is chosen and the required parameters are estimated from the available data. Aspects in data analysis as well as two popular demand distribution functions are presented in the next three subsections.

2.2.1.1 Analysis of Demand Data. A first aspect is to predict future demand as a basis for procurement, production, and distribution planning. A second aspect refers to forecasting errors which cause the necessity of buffers to prevent from a negative performance impact of random noise within the demand forecasts. Quantitative approaches for demand data analysis can roughly be divided into descriptive and explanatory approaches. The descriptive approach applies time series analysis and statistical techniques for exploring demand information, whereas the explanatory approach uses econometric analysis in order to find dependencies between demand realizations and other observable data, for instance price, competitor price, service, quality, economic situation. Though most inventory control models utilize the statistical time series approach, it is necessary to investigate the demand dependency from some general factors beforehand. Otherwise, the pure statistical approach can yield poor performance results though the required information for better planning results would have been predictable with some judgement. If considerable changes in these judgement categories are necessary,

---

[9] Lagodimos, A.G. [1990], p. 74-75.

the demand model must also be reevaluated as to whether it is still appropriate. The most general judgement category concerns the economic situation which affects the consumer behavior to some extent. Connected to this aspect, consumer preferences and influencing activities play an important role for demand realizations. Another aspect that is ignored when applying standard statistical techniques to predict future demand is competitor behavior. The last of other aspects to be mentioned here is the stage of the product life cycle which highly influences the predictability of demands. This aspect has gained increasing attention during the last years. It was observed that demand forecasts using time series methodology for new products and fashion goods with a short product life cycle raises a lot of problems. Therefore, different inventory control and manufacturing concepts have been developed for these applications, this includes quick response and accurate response.[10] The statistical techniques apply well to standard products within their mature phase of the life cycle. In this case, sufficient past demand data is available and the occurrence of dramatical changes is very unlikely. Within the declining life cycle phase, at a minimum, the demand size and the probability distribution characteristics must be adjusted with respect to the market development.

Another pitfall in supply chain inventory control[11] concerns the unsatisfactory coordination between marketing and inventory control. Often, information on marketing activities is not included into inventory decisions. Price changes for the companies or a competitor product, promotional campaigns, and advertising intend to increase sales but these activities fail when they are not coordinated with product availability.

The necessary amount of buffers and related costs are highly influenced by forecasting errors. In order to achieve an optimal operation of the supply chain it is necessary to decrease forecasting errors and demand uncertainty as much as possible. This can be achieved by incorporating general judgement and information from marketing. Another field for forecasting error reduction is offered by the choice of the forecasting model.[12] Nevertheless, some noise will remain and adequate buffers for dealing with the remaining uncertainty are necessary. This problem, although somewhat reduced, will be of central interest in the following discussion.

When concentrating on the time series approach, different demand patterns have to be taken into account. A first important distinction concerns the regularity of the demand pattern. If the demand process faces a large variability, it is called sporadic demand which often occurs in spare parts inventory applications. For consumer goods, there is often a well predictable, regular demand pattern with a small variability. Within a regular demand pattern, three components are important. In order to reduce demand fore-

---

[10] FISHER, M., RAMAN, A. [1996], IYER, A.V., BERGEN, M.E. [1997].

[11] LEE, H.L., BILLINGTON, C. [1992].

[12] For an overview of forecasting techniques in inventory control see SILVER, E.A., PYKE, D.F., PETERSON, R. [1998], p. 74-145.

casting errors to a minimum, it is necessary to filter out these components. The first one is a constant demand level. In this case the forecasting error results from the deviations from this constant level. This is the case modeled in most standard inventory models with stationary, constant parameter demand distributions. In contrast to a base demand level, a trend pattern represents a long run demand increase/decrease, whereas a seasonal demand pattern implies cyclic demand variations from a base level. These effects have to be addressed by inventory models. Otherwise, if only a mean demand and a standard deviation is computed, postulating a base demand level, the forecasting error and the safety stock requirements are larger than necessary. As a consequence, before implementing the base demand level model as it is done in most stochastic inventory models, an analysis of the demand pattern is necessary and in some cases, adjustments of the results are necessary.

An important measure is the forecasting error, especially the standard deviation of the forecasting error. Nevertheless, the inventory control decisions for buffer sizing depend on the complete probability distribution of the forecasting error. Therefore, most inventory control models directly assume a demand probability distribution function. Instead of computing forecasts from demand data, a suitable demand probability distribution function is chosen and the required parameters are fitted from the available data. In order to check whether the chosen demand distribution reflects the customer demand process embedded in the demand data sample, statistical goodness-of-fit tests are available, the most popular is the "Chi square" test.[13] For the application it is important that the parameters of the demand distribution are determined from the data sample. Most probability distributions that are used for inventory control purposes utilize the first two moments in order to characterize the demand distribution. This point is often criticized because only the level and the variability of demand are addressed. An extension is the consideration of the skewness and the kurtosis, related to the third and fourth moment of a demand distribution. Nevertheless, the error made by neglecting this aspect is considerably small.[14] From a sample of $t = 1, ..., n$ past demand observations d_t, the estimators for the demand expectation and the demand variance are

$$\hat{\mu} = \frac{1}{n} \cdot \sum_{t=1}^{n} d_t \tag{2.1}$$

$$\hat{\sigma}^2 = \frac{1}{n-1} \cdot \sum_{t=1}^{n} (d_t - \hat{\mu})^2. \tag{2.2}$$

This probability demand distribution specification implies that the used demand distribution and its parameters are known. This is a further approximation, because in reality the distribution as well as its parameters result from

[13] See SHERBROOKE, C.C. [1992], p. 86-89.
[14] LAU, H.-S., ZAKI A. [1982].

an estimation which underlies an error itself. Nevertheless, the deviation will be small in cases where a large sample of demands is available. In the case of new products with a limited number of available demand realizations, the safety stocks have to be adjusted to incorporate the additional uncertainty about the distribution and the according parameters.[15]

In the following sections, two widely used theoretical demand distributions in the field of inventory control, the normal and the Mixed-Erlang distribution, are presented.[16] Instead of applying a theoretical probability distribution, it is also possible to use the empirical demand distribution given by the sample data. The disadvantage of this approach is the loss of differential analytic calculus which often enables the derivation of solution properties.

2.2.1.2 Normal Distribution. Let the customer demand during a single period be represented by the normally distributed random variable D. The normal distribution $N(\mu, \sigma^2)$ is characterized by two parameters, demand expectation μ and standard deviation σ. The probability density function $\varphi_{\mu,\sigma}(x)$ is given by

$$\varphi_{\mu,\sigma}(x) = \frac{1}{\sigma \cdot \sqrt{2\pi}} \cdot e^{-\frac{(x-\mu)^2}{2\sigma^2}} \tag{2.3}$$

and illustrated in Figure 2.1 for different coefficients of variation $v = \sigma/\mu$. The corresponding cumulative probability density function is

$$\Psi_{\mu,\sigma}(x) = \int_{-\infty}^{x} \varphi_{\mu,\sigma}(x) dx.$$

For the purpose of numerical calculation, the normal distribution $N(\mu, \sigma^2)$ is transformed into the $N(0,1)$ standard normal distribution by substituting $z := \frac{x-\mu}{\sigma}$. The resulting standard normal probability density $\phi_{0,1}(z)$ and the cumulative standard normal density $\Phi_{0,1}(z)$ are given by

$$\phi_{0,1}(z) = \frac{1}{\sqrt{2\pi}} \cdot e^{-\frac{z^2}{2}} \text{ and } \Phi_{0,1}(z) = \int_{-\infty}^{z} \phi_{0,1}(u) du.$$

Modeling customer demands by a normal distribution is only approximate because negative demand values have a positive probability. In order to attain a reasonable approximation, the occurrence probability for this event has to be negligible when using the normal distribution. In Table 2.1 the probability for negative demands $Prob\{D \leq 0\} = \Phi_{0,1}(-\mu/\sigma)$ that depends on the coefficient of variation $v = \sigma/\mu$ is shown. The negative demand probabilities for a large coefficient of variation let a lot of authors suggest to use the normal distribution only if $v \leq 0.5$.[17]

---

[15] RITCHKEN, P.H., SANKAR, R. [1984], SUCHANEK, B. [1996].

[16] References to other theoretical demand distributions used in inventory models are found in SILVER, E.A., PYKE, D.F., PETERSON, R. [1998], p. 272-274.

[17] See e.g. SCHNEIDER, H. [1981], TIJMS, H.C., GROENEVELT, H. [1984].

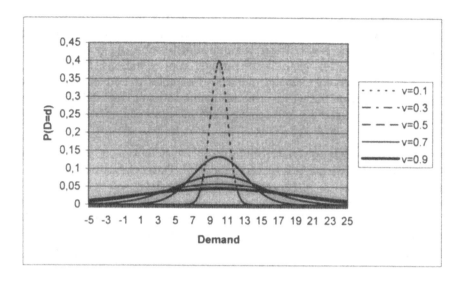

Fig. 2.1. Normal distribution for different coefficients of variation

Table 2.1. Probability for negative demand

v	0.1	0.3	0.5	0.7	0.9
$Prob\{D \le 0\}$	0.0000	0.0004	0.0228	0.0764	0.1335

The normal integral cannot be solved analytically. Therefore, the standard normal distribution is tabulated in order to avoid time consuming numerical integration. Performing inventory control decisions for a large number of products requires fast computation. Therefore, rational approximations for $\Phi_{0,1}(x)$ are proposed.[18]

Several reasons justify the use of a normal distribution for modeling customer demand. It is a well known and widely used probability distribution function. In a number of applications for consumable products, a reasonable goodness of fit is reported. If the total demand of a period is aggregated from a large number of small customers, the central limit theorem that the distribution of a sufficiently large number of probability distributions converges to a normal distribution gives a theoretical foundation. Though the normal integral cannot be solved analytically it is nevertheless often possible to prove properties of an optimal inventory control policy. The disadvantages result from the necessity of rational approximations and from the negative demand problem that makes this distribution unlikely for slow moving item applications where the coefficient of demand variation will, in general, be

---

[18] ABRAMOWITZ, M., STEGUN, I.A. [1970], p. 932, HASTINGS, C. [1955], p. 192. These two references also provide other approximations with different computation time requirements and error terms.

large. Based on this criticism, BURGIN[19] proposed to use the Gamma distribution for inventory control. This continuous distribution is defined only for non-negative values. Nevertheless, the difficulty to solve the gamma integral again results in the use of tables or rational approximations. A mixture of Erlang distributions for demand modeling resembles the advantages of the Gamma distribution but yields a better computational tractability because the cumulative probability and some related integrals can be solved analytically.

2.2.1.3 Mixed-Erlang Distribution. This distribution is suggested by TIJMS[20] and widely used in inventory models.[21] TIJMS claims that any probability distribution can be approximated arbitrarily closely by such a distribution. This generalized Erlang distribution consists of a mixture of pure Erlang distributions. An Erlang distribution results from the sum of $\kappa \in N$ exponentially distributed random variables with the same scale parameter l. The probability density $e_{\kappa,l}$ and the cumulative density $E_{\kappa,l}$ are given by

$$e_{\kappa,l}(x) = \frac{l^\kappa x^{\kappa-1}}{(\kappa-1)!} \cdot e^{-lx}, \qquad E_{\kappa,l}(x) = 1 - \sum_{i=0}^{\kappa-1} \frac{(lx)^i}{i!} \cdot e^{-lx}$$

respectively. Inventory models often apply a mixture of two Erlang distributions with $\kappa - 1$ and κ phases respectively. Then, the underlying demand random variable D is distributed according to an $E_{\kappa-1,l}$ distribution with probability q and according to an $E_{\kappa,l}$ distribution with probability $1 - q$.

$$ME_{\kappa,q,l}(x) = q \cdot E_{\kappa-1,l}(x) + (1-q) \cdot E_{\kappa,l}(x). \tag{2.4}$$

The density of Mixed-Erlang distributions for different coefficients of variation is illustrated in Figure 2.2. TIJMS suggests to apply this demand model for cases with a coefficient of variation smaller or equal to one. For $v > 1$, a hyperexponential distribution is proposed.[22] Because the focus of the following models is in particular directed to standard products with regular demand pattern, the assumption $v \leq 1$ is reasonable and the Mixed-Erlang distribution provides sufficient modeling assistance.

After the expected period demand (μ) and the corresponding standard deviation (σ) have been estimated from some demand data sample by (2.1) and (2.2) respectively, the three distribution parameters κ, l, and q have to be derived. The two-moment fitting procedure outlined by TIJMS determines the parameters such that $E(D) = \mu$ and $E(D^2) - [E(D)]^2 = \sigma^2$ holds for the resulting probability distribution function. Choose κ so that $\kappa^{-1} \leq v^2 < (\kappa - 1)^{-1}$ and

---

[19] BURGIN, T.A. [1975], see also SCHNEIDER, H. [1978].
[20] TIJMS, H.C. [1994], p. 358-359.
[21] DIKS, E.B. [1997], JANSSEN, F.B.S.L.P. [1998].
[22] TIJMS, H.C. [1994], p. 359-360.

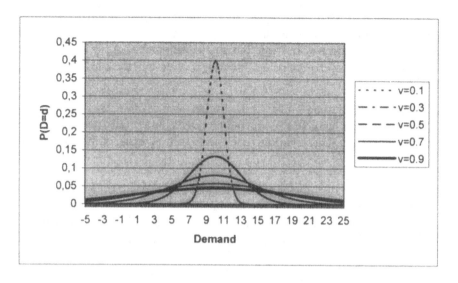

Fig. 2.2. Mixed-Erlang distribution for different coefficients of variation

$$q = \frac{\kappa v^2 - \sqrt{\kappa(1 + v^2) - \kappa^2 v^2}}{1 + v^2}, \qquad l = (\kappa - q)/\mu.$$

The advantage of the generalized Erlang distribution is that it enables one to overcome the shortcomings of the normal distribution, namely the negative demand problem and the computational tractability. Nevertheless, in situations with a small coefficient of variation, numerical problems also arise for this distribution. In this case, the fitting procedure yields a large number of phases κ and the Erlang distribution is computed from a large number of small real values where rounding errors occur more easily. As a consequence, since the normal distribution problem with negative demands becomes negligible in these cases, it can be suggested to use the normal distribution for cases with a coefficient of variation of smaller than 0.5 and the Mixed-Erlang model for values between 0.5 and 1.

2.2.2 Lead Time

The second major influence on the safety stock requirement other than the demand process is the lead time. The lead time is defined as the time span between an order release and the completion of processing or delivery, in other words availability of the products at the stockpoint to satisfy customer demands. When the stockpoint observes demand realizations that have taken place in the last period, the resulting order placed at the supplier is available after the lead time which therefore determines the critical time period where the physical stock cannot be influenced and where protection against uncertain demands is necessary.

In addition to pure processing times, the lead time includes backorder times at the supplier, that is the time span until all required materials are available and processing can start. The single components of the lead time can be classified into order processing, waiting, production, packing, and storage time.

- Order processing time
  This component includes all time requirements for administrative processes at the order releasing stockpoint itself and at the external or internal supply stockpoint.
- Waiting time
  A large fraction of the lead time is imposed by waiting times. These time requirements result from lacking availability of material or from jobs waiting for available processing capacity. If at least one of the suppliers for the required input material is unable to deliver immediately, the backorder time until the complete order is satisfied increases the replenishment lead time of the stockpoint. If the production process takes place on a machine or in a shop that is also utilized by other products, the occupation of this processor induces a further waiting time for free processing capacity.
- Processing time
  When the required material and the processing capacity are available, the pure manufacturing process can start. The process time variability is influenced by the knowledge and training of the workers and also by machine breakdowns. Depending on the detail of the production process modeling, the processing time can also include waiting time components if several elementary working steps are summarized to a single one and waiting times occur in between these elementary units. Another important factor concerns the batch sizes. It is obvious, that the required processing time will depend on the batch size to some extent. Many stochastic inventory models that assume deterministic and constant lead times ignore this aspect.
- Packing and warehouse storage
  After processing, packing and warehouse storage are organized independently from production. The lead time ends with availability of the product for customer demand or requests of succeeding installations. It does not include the transportation time to customers or to succeeding stockpoints, because this is part of the lead time of succeeding processes.

In inventory control models, one main assumption concerns the grid of lead time values. Periodic review models, where the inventory status is inspected and replenishment decisions are made every multiple of a base period, often assume that the lead time is a multiple of the base period and therefore integer valued. Nevertheless, the outcome of such a model is always being influenced by the definition of the base period, e.g. one hour, an eight-hour shift, a day, or a week. Therefore, a modeling of lead times by real numbers even in periodic review inventory control models is reasonable. Besides events occurring within a base period, real valued lead times can alternatively be interpreted as long run average values.

Another assumption is made concerning the knowledge about the lead time. One implication of a deterministic lead time assumption in a single-echelon modeling context is that the supplier can always deliver any requested amount of products within this time span. Models with variable lead time can be seen as a model extension. The most simple way of this variability extension is to model stochastic lead times characterized by expected value and standard deviation. More elaborate models explain some part of lead time variability, especially the waiting time component. KARMARKAR[23] reviews queuing approaches to model the relation between lead times, lot-sizes, and capacities. Nevertheless, shop floor decisions at the supplier can be neither influenced nor observed by the stockpoint in general. This requires for modeling lead time as a random variable in some applications.

Demand over the replenishment lead time determines the relevant planning information for safety stock decisions. The following subsections present how to determine lead time demand. The cases of deterministic and stochastic lead times are treated independently. Further problems arise if the demand process is correlated in time which is addressed in 2.2.2.3.

2.2.2.1 Deterministic Lead Time. Assume a given single period demand model without correlation. For a deterministic lead time λ, the probability distribution of cumulative demand within this time span results from the λ-fold convolution of the single period demand distribution.

- Normal distribution
  If single period demand is modeled by a normal distribution $N(\mu, \sigma^2)$, the lead time demand distribution is also normally distributed with expected value $\mu \cdot \lambda$ and standard deviation $\sigma \cdot \sqrt{\lambda}$.

$$\varphi_\lambda = N(\mu\lambda, \sigma^2\lambda)$$

- Mixed-Erlang distribution
  If single period demand is modeled by a mixture (with probabilities q and $1-q$ respectively) of $E_{\kappa-1,l}$ and $E_{\kappa,l}$, the exact lead time demand distribution consists of a mixture of $\lambda+1$ Erlang distributions $E_{(\kappa-1)\cdot\lambda,l}, ..., E_{\kappa\cdot\lambda,l}$. The probability density function is given by[24]

$$\varphi_\lambda(x) = \sum_{i=0}^{\lambda} \binom{\lambda}{i} q^{\lambda-i}(1-q)^i \cdot e_{(\kappa-1)\lambda+i,l}(x).$$

In order to overcome the additional computational effort to evaluate $\lambda+1$ Erlang distributions, SEIDEL, DE KOK[25] suggest to approximate the lead

[23] KARMARKAR, U.S. [1993].

[24] VAN HOUTUM, G.J., ZIJM, W.H.M. [1997].

[25] SEIDEL, H.P., DE KOK, A.G. [1990].

time demand by a mixture of two Erlang distributions $E_{\kappa-1,l}$, $E_{\kappa,l}$. The distribution parameters are computed by the same fitting procedure as outlined for the single period demand model, only the lead time demand characteristics expected value $\lambda \cdot \mu$ and standard deviation $\sqrt{\lambda} \cdot \sigma$ are used instead of μ and σ.

2.2.2.2 Stochastic Lead Time. In the stochastic lead time case, the resulting demand distribution is in general not of a simple type even if a simple standard distribution is assumed for single period demands. Therefore, an approximation idea similar to the one used by SEIDEL, DE KOK is suggested by TIJMS, GROENEVELT.[26] A theoretical demand distribution such as the normal or the Mixed-Erlang one is used and the parameters are fitted on the first two moments of lead time demand. Let μ and σ_D^2 denote single period demand expectation and variance respectively and λ and σ_λ^2 denote lead time expectation and variance respectively. Under the assumption that the random variables for customer demand and lead time are independent and that orders do not cross, i.e. the orders arrive in the sequence they were released, the expected value and the variance of cumulative demand during the lead time are given by

$$E[D(\tilde{\lambda})] = \lambda\mu \qquad (2.5)$$
$$Var[D(\tilde{\lambda})] = \mu^2\sigma_\lambda^2 + \lambda\sigma_D^2. \qquad (2.6)$$

The advantage of this modeling approach is the reduction of the problem to a deterministic lead time problem with adjusted demand characteristics to account for the lead time uncertainty. On the other hand, the quality of the methods outcome depends on the similarity of the true lead time distribution to the assumed model, e.g. normal or Mixed-Erlang. EPPEN, MARTIN[27] present examples that this approach can completely fail in specific situations and suggest an improved approach.

An alternative approach exists that reduces the stochastic demand, stochastic lead time problem to a stochastic demand, deterministic lead time problem. This approach is especially popular within the MRP framework. A planning lead time $\hat{\lambda}$ is determined for inventory control purposes. The required material is ordered $\hat{\lambda}$ periods before it is needed. The planning lead time consists of the lead time expectation λ and a safety lead time increment for lead time variations. This safety increment is assumed to be a multiple of the lead time standard deviation σ_λ.

$$\hat{\lambda} = \lambda + k \cdot \sigma_\lambda. \qquad (2.7)$$

Similar to the safety stock planning approaches to be presented in the next section, one possibility to fix an appropriate safety lead time is to set the

[26] TIJMS, H.C., GROENEVELT, H. [1984].
[27] EPPEN, G.D., MARTIN, R.K. [1988].

safety surplus in such a way that the probability, that the lead time random variable $\tilde{\lambda}$ does not exceed the planning lead time is larger or equal to some predetermined fraction α_λ, i.e. α_λ percent of all replenishment orders arrive within the planning lead time.

$$Prob\{\tilde{\lambda} \leq \hat{\lambda}\} \geq \alpha_\lambda$$

The additional incorporation of timing uncertainty requires an analysis of lead time data. Similar techniques as discussed for stochastic demand data analysis apply and similar special problems occur. Expected value and standard deviation can be estimated from a lead time data sample by formulas (2.1) and (2.2) respectively. By a formulation of more elaborate models that explain some of the interrelation of lead times, lot-sizes, and capacities, more variation can be explained and, therefore, external (unexplained) lead time variation and buffer requirements are reduced. Again, different patterns for stochastic lead times can occur. A trend pattern in lead time can be explained by learning curve effects or by dynamic improvement of manufacturing technology. Seasonal lead time effects can result from capacity utilization depending on multiple products with seasonal demand patterns.

2.2.2.3 Demand Correlation. In the previous sections it was assumed that demands are independent of the demand history. Nevertheless, in some practical situations a dependency between demands of different time periods exists. A positive absolute variation can follow a negative variation if customers postpone or anticipate their requirements. In these cases, the additional sales are connected with lower sales in the past or in the future and demands are negatively correlated with some time lag. Positive correlation between demands is given if positive variations follow positive variations or negative variations follow negative ones. This additional dependency influences the safety stock requirements over the replenishment lead time. If lead time demand as the sum of the single period demand realizations over consecutive time periods is positively correlated, the variability of lead time demand is larger and more safety stocks are required. On the other hand, negative correlation implies risk diversification within the lead time and safety stock requirements reduce.

The following safety stock considerations are similar to the previous sections. Mean and standard deviation of lead time demand are derived following the approach of CHARNES, MARMORSTEIN, ZINN[28] for a deterministic lead time. This approach does not require the assumption that demands are generated from a parametric time series process. For stochastic lead times only an idea for possible extensions is considered.

Let $g(i) = E[(D_t - \mu)(D_{t-i} - \mu)]$ denote the autocovariance of demand at lag i. Then, the variance of demand during the lead time is

---

[28] CHARNES, J.M., MARMORSTEIN, H., ZINN, W. [1995].

$$\sigma^2 = (\lambda + 1) \cdot \sigma_D^2 + \sum_{i=1}^{\lambda} (\lambda + 1 - i) \cdot g(i).$$

The required autocovariance coefficients can be estimated from a demand data sample of n observations by

$$\hat{g}(i) = \frac{1}{n} \sum_{t=1}^{n-i} (D_t - \mu)(D_{t+i} - \mu), \qquad i = 0, ..., \lambda.$$

For stochastic lead times, CHARNES, MARMORSTEIN, ZINN suggest to use the same approach with $\lambda = |E[\tilde{\lambda}]| + 1$ for samples with a small and $\lambda = |E[\tilde{\lambda}]| + m$ for larger coefficients of lead time variation. m denotes some positive integer increment to the expected lead time. This approach is similar to the use of a safety lead time. A more theoretical approach is presented by RAY[29].

The above approach is independent from any time series model specification. For the following two popular first lag time series models, moving average MA(1) and autoregressive AR(1) processes, the autocovariance coefficients can be specified as follows:

- First order moving average process MA(1):

$$D_t = \epsilon_t - b \cdot \epsilon_{t-1}, \ -1 \le b \le 1, \epsilon_t \sim N(0, \sigma^2)$$
$$g(0) = \sigma^2, g(1) = -b, g(i) = 0, i = 2, 3, ...$$

- First order autoregressive process AR(1):

$$D_t = aD_{t-1} + \epsilon_t, |a| < 1, \epsilon_t \sim N(0, \sigma^2)$$
$$g(0) = \sigma^2, g(i) = a^i \cdot g(0), i = 1, 2, ...$$

Under the popular practice to estimate demands by an exponential smoothing forecasting technique (with a smoothing constant a), forecast errors become correlated even when the underlying demand process is not correlated. The resulting variance of forecast errors in relation to the inherent demand variance is shown to be[30]

$$\sigma^2 = \frac{2}{2-a} \cdot \sigma_D^2.$$

The variance of forecast errors for which safety stocks are necessary is larger $(2/(2-a) > 1)$ than the pure demand variance.

---

[29] RAY, W.D. [1980], [1981]
[30] EPPEN, G.D., MARTIN, R.K. [1988].

2.2.3 Input Coefficients and Yield

The material input coefficients are the second component that characterize the input process. Input coefficients $a_{j,i}$ denote the amount of material j (measured in the physical dimension of j) that is required to produce or transform this material into one unit of the product assigned to stockpoint i. This characterization implies that the underlying process is output quantity driven and if alternative activities to manufacture the product are available, one single recipe has been predetermined. Every process yields a single (desired) output and the required input solely depends on the target output quantity. By-products that occur with some chemical processes are excluded from the following modeling. The input coefficient includes all material, also the material that results in scrap and by-products. Depending on the reliability of the process and the variability of the material requirements, the input material needs can be modeled constant and deterministic or stochastic.

In cases with a negligible variability (uncertainty) in yield, it is reasonable to assume deterministic input coefficients $a_{j,i}$. The demand characteristics of input material j are derived from the output requirements for product i. In the stochastic modeling variant, the problem of random yield is rather complicated.[31] In order to incorporate process yield variability in a simple way, the approach presented by WIJNGAARD, WORTMANN[32] can be applied. The idea behind their approach is similar to the simple incorporation of stochastic lead times. Let $\tilde{r}$ define a random process level necessary to obtain a usable yield of one item. Let r and σ_r^2 denote the respective expected value and variance. The coefficients $a_{j,i}$ are assumed to be proportional to the level r. Under the additional assumption that demand, lead time, and yield are independent random variables and orders do not cross, the expected lead time requirements and the corresponding variance are given by

$$E[D(\tilde{\lambda})] = \lambda \mu r \qquad (2.8)$$
$$Var[D(\tilde{\lambda})] = \lambda \sigma_D^2 + \mu^2 \sigma_\lambda^2 + \sigma_r^2. \qquad (2.9)$$

The extension of this approach to applications with rework options for defective products may become critical due to the correlation of lead time and input requirements.

Within deterministic planning environments like MRP, the random requirements level is transformed into a planning level in the same manner as proposed for random lead times.

$$r_p = r + k \cdot \sigma_r$$
$$Prob\{\tilde{r} \leq r_p\} \geq \alpha.$$

---

[31] See YANO, C.A., LEE, H.L. [1995].
[32] WIJNGAARD, J., WORTMANN, J.C. [1985].

2.2.4 Inventory Control and System Dynamics

Target of inventory control is to coordinate output and input processes in order to avoid too large physical stocks and to prevent stockouts connected with poor customer service. Inventory control rules give advice when and how much to order at the supplier or at the preceding installations. Concerning the timing when replenishment decisions are made, periodic and continuous review inventory control policies are distinguished.

Under a periodic review policy, the stockpoint's inventory state is inspected once every R periods (be it hours, days, weeks or even months) and an inventory control rule is applied to determine if and how much products have to be ordered. Under a continuous review policy, the inventory state is inspected after every single demand event. Therefore, periodic review policies are often used for fast moving standard items. For slow moving and especially high valued products (e.g. spare parts) continuous review policies are more appropriate. In the following, it is assumed that the review period is equal to one ($R = 1$), i.e. the model is formulated in a way that the base period is identical to the review period.

Before, two special inventory control rules are introduced and the system dynamics of state variables that depend on demand realization and replenishment decisions are outlined. For simplicity of presentation, the case of an integer and deterministic lead time λ is considered. For the customer behavior in stockout situations, the backorder and the lost sales case are addressed. The backorder case is presented according to a FIFO delivery rule where the oldest backlog is served first.

The net stock y_t at the end of period t results from the net stock y_{t-1} at the end of the previous period plus the order quantity $x_{t-\lambda}$ that arrives at the beginning of period t, and therefore was released at the beginning of period $t - \lambda$, minus the demand d_t that occurred during period t.

$$y_t = y_{t-1} + x_{t-\lambda} - d_t.$$

In the lost sales case, the net stock never becomes negative. If the initial stock plus the arrived order are not sufficient to meet the demand during the corresponding period, excess demand is lost and the net stock is zero.

$$y_t = \max\{0; y_{t-1} + x_{t-\lambda} - d_t\}.$$

The on-hand stock OH_t at the end of period t represents the available physical stock and is identical to the net stock in the lost sales case.

$$OH_t = \max\{0, y_t\}$$

The backlog BL_t is the amount of stock that has been requested by customers but has not been delivered due to stock insufficiencies at the end of period t. In the backorder case where the sum of on-hand stock and backlog is identical to the net stock

$$BL_t = \max\{0, -y_t\}$$

holds whereas in the lost sales case the backlog is identical to zero by definition ($BL_t \equiv 0$). The shortage is defined as the period demand that could not be satisfied from available inventory. It is given by the excess of demand over the available inventory (after partly or completely filling the backorders of the past period).

$$SH_t = \max\{0, d_t - \max\{0, y_{t-1} + x_{t-\lambda}\}\}$$

The relevant inventory state information for replenishment decisions at the beginning of period t is given by the inventory position IP_t. This figure is measured at the beginning of period t after the arrival of the order released λ periods ago but before the order release in t. It includes all orders of previous periods that have not arrived at the stockpoint and represents the amount of stock that cannot be influenced by decisions of the stockpoint.

$$IP_t = y_{t-1} + \sum_{i=1}^{\lambda} x_{t-i}.$$

In the following, ordering decisions x_t are derived from standard periodic review inventory control rules. Two different inventory control rules that are related to lot-for-lot and batch ordering respectively are presented.

- (S)-policy
  For an (S)-inventory control rule, the replenishment order size is chosen in way that the inventory position after ordering becomes equal to the order-up-to-level S. Therefore, the inventory position is increased to a constant base level every review period.

$$x_t = \max\{0, S - IP_t\}$$

- (s,S)-policy
  When replenishment decisions are determined from an (s, S) rule, the order release is conditioned. Only if the inventory position reaches a level below or equal to the reorder point s, a replenishment order will be placed that increases the inventory position to the order-up-to-level S. This policy becomes advantageous if a setup cost is present and therefore reordering every review period is too expensive.

$$x_t = \begin{cases} S - IP_t & \text{if } IP_t \leq s \\ 0 & \text{otherwise} \end{cases}.$$

An introductory overview on different control rules is given by SILVER, PYKE, PETERSON and HADLEY, WHITIN.[33] Under stationary conditions and

---

[33] SILVER, E.A., PYKE, D.F., PETERSON, R. [1998], HADLEY, G., WHITIN, T.M. [1963].

the backorder case without setup costs the (S) policy is optimal.[34] Under the presence of setup costs the (s, S) policy has been proven to be optimal.[35] In the lost sales case, the structure of an optimal policy is rather complex.[36] Nevertheless, the presented policies can be used in the lost sales case as an approximation if the parameters are adjusted accordingly.

2.2.5 Cost and Service Performance Measures

The quality of an inventory control policy can be evaluated with respect to quantity and cost measures. Quantity performance concerns internal operational performance that is measured by inventory levels and external performance to customers measured alternatively by shortage or backlog characteristics. These considerations are made with respect to average measures due to the long run repetitive nature of the decision problem. In the following, relationships for the determination of different performance measures are outlined and specified for the two control policies that were introduced in the previous section.

The average physical stock can be evaluated at several points of time. Inventory literature provides models where stocks are evaluated at the beginning or at the end of a period or even where the exact inventory trajectory within a single period is modeled explicitly. For ease of presentation, the following discussion is restricted to the most common assumption of an evaluation at the end of a period. Given the inventory position IP_t at the beginning of a period after a replenishment decision, the expected on-hand stock $E[OH(IP_t)]$ is given by

$$E[OH(IP)] = \int_0^{IP} (IP - x)dF_{\lambda+1}(x).$$

Under lot-for-lot replenishment, the inventory position is increased to the order-up-to-level every period and the average on-hand stock is given by the above expression with $IP = S$. When the inventory is controlled by a periodic review reorder-point/reorder-level policy, the steady state distribution of the inventory position is required to evaluate the performance measures. The inventory position density is given by[37]

$$f(x) = \frac{m(S - x)}{1 + M(S - s)}, \qquad x \le S$$

where $M(\cdot)$, $m(\cdot)$ are the renewal function and its density.[38] Then, the expected on-hand stock for an (s, S) policy becomes

---

[34] BELLMAN, R.E., GLICKSBERG, I., GROSS, O. [1955].
[35] SCARF, H. [1960].
[36] NAHMIAS, S. [1979].
[37] KARLIN, S. [1958].
[38] IGLEHART, D.L. [1963].

$$E[OH] = \frac{E[OH(S)] + \int_0^{S-s} E[OH(S-x)]m(x)dx}{1 + M(S-s)}$$

by averaging the conditional on-hand stocks with the inventory position steady state probabilities. Safety stock is a performance measure similar to the on-hand inventory. In contrast to the physical stock, the safety stock is defined as the average net inventory at the end of an arbitrary period and therefore includes all events with a positive backlog at the end of a period. Conditioned on the inventory position, the safety stock is

$$SS(IP) = \int_0^{\infty} (IP - x)dF_{\lambda+1}(x).$$

The average absolute safety stock can be derived analogously to the average physical stock. Note, that $SS = E[OH] - E[BL]$. If the order-up-to-level is sufficiently large and therefore the average backlog is negligible small, the on-hand stock can be approximated by the safety stock. Conditioned on the inventory position, the average backlog at the end of a period is

$$E[BL(IP)] = \int_{IP}^{\infty} (x - IP)dF_{\lambda+1}(x).$$

As pointed out for an (S)-policy, the average backlog results from $E[BL] = E[BL(S)]$. For an (s, S)-policy, the average backlog is given by

$$E[BL] = \frac{E[BL(S)] + \int_0^{S-s} E[BL(S-x)]m(x)dx}{1 + M(S-s)}.$$

A shortage only includes the newly generated backorders of a period. Given the inventory position, the expected shortage is

$$E[SH(IP)] = \int_{IP}^{\infty} (x - IP)dF_{\lambda+1}(x) - \int_{IP}^{\infty} (x - IP)dF_{\lambda}(x)dx.$$

The required performance measures for the two different inventory policies again result from setting $IP = S$ in case of lot-for-lot ordering and by averaging the conditional expressions with the steady state probabilities for the inventory position in case of batch ordering.

$$E[SH] = \frac{E[SH(S)] + \int_0^{S-s} E[SH(S-x)]m(x)dx}{1 + M(S-s)}$$

The difficulty in evaluating the above expressions arises from the derivation of the renewal quantities. Therefore, approximations have been derived.[39]

For economic decision and accounting purposes, it is necessary to value these performance quantities with costs. With respect to the repetitive and

---

[39] SCHNEIDER, H. [1978].

dynamic nature of an inventory control system, the literature distinguishes between expected average cost and expected discounted cost approaches. The remainder of this work applies an average cost framework. The objective is to minimize the sum of ordering, holding, and shortage costs. As it is common in most business applications, the performance quantities are valued with linear purchase, production, holding, and stockout penalty cost rates respectively. Extensions to convex cost functions to include capacity aspects or concave cost functions in order to model economies of scale complicate the analysis. Therefore, they are omitted to keep the analysis and presentation simple.

(1) Ordering and production cost

This cost category can be divided into a fix (K per order) and a variable (c per unit) component. In an (S)-policy where fixed costs are assumed to be negligible ($K \approx 0$) the average variable operating costs per period are $c\mu$. Systems that include fixed costs, incorporated by applying an (s, S) policy, provide average operating costs of

$$\frac{K + c \cdot (S - s + \frac{\mu_2}{2\mu})}{1 + M(S - s)}.$$

The quantity multiplied by c represents the average order quantity in a periodic review (s, S) system which is given by the minimum order quantity $S - s$ plus the expected undershot of the reorder level. μ_2 and μ represent the second and first moments of the demand distribution respectively.

(2) Holding cost

The class of holding cost is divided into two subcategories, out-of-pocket cost that are connected with direct payments and opportunity costs for capital investment. The quantity valued with a linear holding cost rate per item and period is in general the on-hand stock at the end of a period or, as an approximation, the safety stock.

(3) Stockout penalty cost

Costs induced by stockouts consists of out-of-pocket cost and opportunity cost which include costs of delayed payments as well as goodwill losses. Depending on the cost category, these values occur once (with the occurrence of a shortage) or every period where the shortage remains unsatisfied. Therefore, the backlog as well as the new shortage in a period can represent the basis for stockout costs.

In a pure cost model, all cost components are aggregated and the goal is to find a policy (or policy parameters for a given type of inventory control rule) that minimizes total expected costs. Because of accounting difficulties in estimating shortage costs, the performance according to the output process is alternatively evaluated by service measures. If some predefined service in the sense of one of the below service measures is guaranteed to the customers, the objective of the inventory system reduces to the minimization of holding cost subject to a service level constraint.

Service level definitions can be based on different time intervals:[40] on an average period, on an average replenishment cycle, or according to the lead time. Service levels based on the order cycle have the advantage of simplifying computation. From the customer point of view, this service measure has the shortcoming that the length of the order cycle of the stockpoint is of minor interest, whereas the quality of demand satisfaction in every period is the relevant performance indicator. Therefore, only period-based service level definitions are considered in the following. Three definitions of service levels in inventory models are discussed[41] that differ with respect to their penalization of shortages. In some situations, only the occurrence but not the magnitude of a shortage is of relevance (α-service level). In other applications, the magnitude of a shortage in an arbitrary period (β-service level) or the backlog (γ-service level) which additionally incorporates old, still unsatisfied, shortages is considered. It is obvious that these service level definitions are related to the stockout penalty cost categories.[42]

In case of a stockout occurrence related measure, service is defined as the relative number of periods without shortage. In the inventory control model, α is defined as the non-stockout probability, i.e the probability that net inventory at the end of an arbitrary period is non-negative.

$$\alpha = Prob\{y \geq 0\} \qquad (2.10)$$

The stockout penalization behind this service definition is related to the occurrence of problems (independent of their size). The stockout magnitude related service measure β is defined as the fraction of demand that is immediately delivered from stock.

$$\beta = \frac{\text{mean satisfied demand per unit of time}}{\text{mean demand per unit of time}}$$

In terms of unsatisfied demand, this expression can be rewritten to

$$\beta = 1 - \frac{\text{mean unsatisfied demand per unit of time}}{\text{mean demand per unit of time}}. \qquad (2.11)$$

The β-service level (also called "fill rate") measures the quantity of new shortages created in an arbitrary period in relation to the total demand of a period. This measure still does not account for the duration of a stockout. The backlog related service measure γ incorporates this criterion by measuring the entire backlog instead of new shortages. Note that this criterion is only relevant in the backorder case since the lost sales case excludes the accumulation of backorders.

---

[40] SCHNEIDER, H. [1981].

[41] SCHNEIDER, H. [1981].

[42] ALSCHER, J., SCHNEIDER, H. [1982], SILVER, E.A., PYKE, D.F., PETERSON, R. [1998].

$$\gamma = 1 - \frac{\text{mean cumulative unsatisfied demand per unit of time}}{\text{mean demand per unit of time}} \qquad (2.12)$$

In contrast to quantity related service measure definitions, a second stream uses time criteria to measure inventory control performance. TEMPELMEIER analyzes an inventory model with a service constraint on the expected customer order waiting time.[43]

The empirical measurement of the service measure definitions is illustrated by the following example. The performance of an (S)-policy with an order-up-to-level of 100 units is considered over a horizon of 8 periods. The initial state of the system consists of an on-hand stock of 100 units, zero outstanding orders, and no backorders. The replenishment lead time is $\lambda = 2$ periods.

Table 2.2. System dynamics in the service level example

Period	0	1	2	3	4	5	6	7	8
d_t	-	50	40	70	30	10	40	50	30
OH_t	100	50	10	0	0	0	20	0	0
SH_t	0	0	0	60	30	10	0	0	0
BL_t	0	0	0	60	40	10	0	0	0

Using the definitions introduced above, the realized empirical service levels are $\hat{\alpha} = 62.5\%$, $\hat{\beta} = 68.75\%$, and $\hat{\gamma} = 65.63\%$.

SILVER, PYKE, PETERSON[44] summarize factors that influence the choice between the different service measures and the determination of their size. This decision can differ from product to product and is influenced by market competition, customer preferences, their behavior in stockout situations, the degree of substitution possibilities, and on the availability of activities to replenish the product from alternative sources. Nevertheless, the choice of type and size of a service level constraint remains a decision to be based on managerial intuition. Compared to the implementation of a stockout cost penalty framework, the service level approach offers some management accounting advantages. The observed service fulfillment can easily be compared with a predetermined service target. This advantage becomes even more obvious for supply chain coordination problems. The overall supply chain performance is highly dependent on all involved operations, whereas stockout penalties are only detected at the final level. In order to achieve a required customer service target, internal supply chain agents contribute by fulfilling some internal service degrees. This method offers more transparency than allocating stockout costs over the supply chain in order to control materials coordination. In addition, observed service level deviations can indicate long run changes in

---

[43] TEMPELMEIER, H. [1985].
[44] SILVER, E.A., PYKE, D.F., PETERSON, R. [1998].

the systems operating environment, for instance a change in market competition or product life cycles. These investigations into model and parameter validity can reveal necessary changes in the implemented control parameters or even in the applied control rule itself.

2.3 Safety Stock Planning Under Service Level Constraints

In this section, the safety stock determination for a single stockpoint is outlined. Formulas and algorithms are summarized for lot-for-lot and batch ordering policies, and additionally for normally and Mixed-Erlang distributed demands. These methods recommend theoretically satisfactory safety stock levels that depend on the main influencing factors of the inventory system. In contrast to these theoretical models, a lot of "rules of thumb" for safety stock determination can be found in the literature and dominate in practice[45]. This argumentation is supported by the observation that unsatisfactory safety stock planning, especially the independence of safety stock size and the degree of demand or lead time variability, is a frequently detected pitfall in supply chain management.[46] The following material summarizes some of these rules.

- Multiple of the expected period demand
  This rule relates to demand uncertainty driven safety stock requirements. The safety stock level is set, for example, equal to the expected demand of 4 weeks. From a theoretical point of view, the shortcoming of this approach is that it prescribes safety stock levels that are independent of demand variability.
- Multiple of periods supply
  This rule is related to supply uncertainty and implements stock levels to cover delays in supply. Nevertheless, this approach also does not address supply variability.
- Maximum reasonable demand during the lead time
  Another approach is based on the idea to provide instantaneous service in all reasonable demand situations. The resulting safety stock norm covers the maximum demand variability during the replenishment lead time.

The previous three approaches to solve the safety stock planning problem are mainly based on practical intuition. SILVER, PYKE, PETERSON present the following criteria for establishing safety stocks that can be regarded as more theory based.

---

[45] In an empirical study, EL-NAJDAWI [1993] found that none of the responding firms applied the square root safety stock formula.

[46] LEE, H.L., BILLINGTON, C. [1992].

- Square root safety stock formula with standard safety factor
  Within the square root approach, the safety stock depends on the standard deviation of lead time demand multiplied by a constant factor k (e.g. $k = 2$) and therefore is directly connected to demand and lead time variability. As outlined in the demand modeling section, the lead time demand variability is given by the single period demand multiplied by the square root of the lead time length. Nevertheless, the safety factor that reflects the cost consequences of stockouts or service performance remains to be determined. In order to avoid the problems of measuring stockout penalty cost rates or defining a service level constraint this approach uses a common standard factor. This safety factor choice is based on intuition (as criticized for service measure and level assignment). In addition, a service level is more informative than a safety factor.
- Safety stock according to minimization of stockout costs
  The determination of the impact of economic criteria on safety stocks in the previous approaches has been solved by intuition. A theoretically satisfactory solution has to be based on the optimization of economic performance trade-offs. As outlined in Section 2.2.5, one possibility is to follow a pure cost approach and to assign holding costs to inventories and penalty costs to backorders and shortages. The trade-off between costs of excess inventory and penalty costs yields an optimal safety stock level.
- Safety stock according to service level constraints
  The alternative performance measure service level used to avoid penalty cost measurement problems prescribes a safety stock level determined by minimizing inventory holding costs subject to a service level constraint. Because of the monotonicity of the safety stock level (with respect to the service level), the resulting safety stock will be the inventory level that satisfies the constraint as an equality. Compared to the cost approach, an equivalence property holds in the sense that for each service level constraint, a corresponding stockout penalty cost parameter exists that yields the same safety stock level.
- Safety stock planning subject to aggregate constraints
  In contrast to planning safety stocks for individual products independently, overall operational and budget constraints may require a simultaneous consideration. These approaches are characterized by a joint financial budget for inventories and by some aggregate definition of system service on the customer side. The economic criterion to be followed can either be to maximize aggregate system service performance with respect to a given budget or to achieve a desired level of system performance with minimal consumption of budget resources.[47]

---

[47] A general description is found in SILVER, E.A., PYKE, D.F., PETERSON, R. [1998]. LAGODIMOS, A.G. [1990] discusses the aggregate safety stock problem for a multi echelon situation whereas SHERBROOKE, C.C. [1992] addresses a multi-echelon system with emphasis on spare parts and repair operations.

These approaches for safety stock planning divide into explicit and implicit methods. Explicit methods directly focus on the prescription of a buffer level, e.g. setting a desired multiple of periods demand or supplies or fixing the safety factor to some level. The investigation of inventory control rules like (S)- or (s, S)-policies with respect to optimal or satisfactory policy parameters implicitly defines a safety stock level. One special problem arises when explicit and implicit methods are compared: the problem of a negative planned safety stock level. Within the implicit approach, negative safety stocks can occur in situations where stock holding is rather expensive or if other benefits like risk pooling can be drawn from postponing demand satisfaction. In these cases, inventory holding cost can be saved by implementing a negative expected inventory level, i.e. a planned backlog, without violating the service level constraint. SNYDER[48] recommends negative safety stocks for systems with significantly large lot-sizes. Then, a considerable amount of service is already achieved by the availability of cycle stocks. From a cost minimization point of view, it is optimal to order when already some backlog has been accumulated. In multi-stage distribution systems, a negative safety stock level is often recommended for the central depot stockpoint because of increased risk pooling. The desired service levels at the local warehouse can be achieved by delayed allocation of deliveries from the central depot and, therefore, no additional safety stock is necessary.

On the other hand, explicit safety stock planning approaches almost exclusively assume non-negativity of safety stock levels. For approaches like fixing the safety stock level to a multiple of periods supply or demand, this assumption is fulfilled implicitly whereas other models add a non-negativity constraint.[49] Since a negative planned safety stock implies a planned delivery delay, HAX, CANDEA[50] conclude a service level misspecification for customer side stockpoints and suggest an adjustment of the underlying service level. A negative safety stock level at an internal stockpoint that supplies other installations of the supply chain generates planned internal delivery disturbances that are often undesired from an organizational point of view. Based on these reasons, the following analysis of single-echelon as well as the main part on multi-echelon safety stock planning assumes that planned safety stock levels cannot be negative.

Assumption 1 *The planned safety stock level is non-negative, i.e. $k \geq 0$.*

Necessary adjustments in cases where a negative safety stock level would occur are discussed in the following section together with the inventory control rules.

---

[48] SNYDER, R.D. [1980].

[49] See SIMPSON, V.P. [1976], DAS, C. [1978].

[50] See HAX, A.C., CANDEA, D. [1984], p. 198.

2.3.1 Lot-for-Lot Ordering

In situations where fixed ordering costs are zero or negligible, lot-for-lot ordering is a reasonable replenishment strategy. If demand is stochastic, an order-up-to-(S) policy implements such a strategy where the ordering quantity at the beginning of each period equals the realization of the demand random variable of the previous period. For a given type and level of service measure, the inventory control policy parameter S is determined in a way that the required target service level is attained. Therefore, it is necessary to derive analytic expressions for the relation between the order-up-to-level S and the resulting service level. First, the required relationships between the order-up-to-level and the three service performance measures introduced in 2.2.5 are presented for an arbitrary type of demand distribution.

- α-service level
  The probability of a positive backlog at the end of an arbitrary period is constrained by the level $1 - \alpha$. Since the inventory position equals S after each replenishment decision, the probability of a positive backlog in an arbitrary period is equal to the probability that the cumulative demand of $\lambda + 1$ periods exceeds S. An overshot of service level fulfillment implies larger safety stocks and therefore, larger holding costs than necessary. As a consequence, the constraint will always hold as an equality. An equivalent formulation of the service level constraint requires that the probability that cumulative demand of $\lambda + 1$ periods does not exceed S is equal to α.

  $$F_{\lambda+1}(S) = \alpha \qquad (2.13)$$

  In order to illustrate the relationship between the service level constraint and the stockout penalty cost approach, consider the infinite horizon newsboy problem.[51] The objective is to minimize average expected holding and backordering cost. A penalty p is incurred per item short per period and holding costs h are assigned to physical stocks per item and period. The stationary order-up-to-level that solves the trade-off between holding safety inventory and backorder penalty costs results from

  $$F_{\lambda+1}(S) = \frac{p}{p+h}.$$

  Therefore, from a cost point of view, the optimal α-service level that yields the equivalence between both approaches is equal to the fraction of p and $p + h$.

- γ-service level
  A γ-type service level constraint limits the expected backlog at the end of an arbitrary period. Under an (S)-inventory control rule, the expected

---

[51] See e.g. LEE, H.L., NAHMIAS, S. [1993], p. 26-28.

backlog is defined as the cumulative unsatisfied customer demand. It is derived from the partial expectation over all cumulative demand realizations over $\lambda + 1$ periods that exceed the order-up-to-level.

$$\int_S^\infty (x - S)f_{\lambda+1}(x)dx = (1 - \gamma)\mu \tag{2.14}$$

- β-service level
  In contrast, a β-type constraint limits the expected shortage during an arbitrary period. The shortage is computed as the expected cumulative backlog of $\lambda + 1$ periods (at the end of an arbitrary period) minus the cumulative expected backlog over λ periods (the already existing backlog at the beginning of an arbitrary period).

$$\int_S^\infty (x - S)f_{\lambda+1}(x)dx - \int_S^\infty (x - S)f_\lambda(x)dx = (1 - \beta)\mu \tag{2.15}$$

These results serve as a basis for the determination of order-up-to-levels under normally (recommended for situations with a small coefficient of variation) and Mixed-Erlang distributed demands (for large demand variability scenarios). The three service level constraint equations implicitly define the required order-up-to-level. However, an analytical solution with respect to S is not possible. Therefore, numerical search techniques and approximations are proposed in the literature. Two common iterative procedures to find a solution of implicitly defined functions, bisection and the Newton-Raphson method, can be found in Appendix C.

Assume that the single period demand random variable is normally distributed $D \sim N(\mu, \sigma^2)$. The normally distributed lead time demand over $\lambda + 1$ periods $N((\lambda + 1)\mu, (\lambda + 1)\sigma^2)$ can be transformed into a $(0, 1)$ normal distribution by the standardization $k = (x - (\lambda+1)\mu)/(\sigma\sqrt{\lambda + 1})$. Using this concept, the order-up-to-level becomes

$$S = \mu \cdot (\lambda + 1) + k \cdot \sigma \cdot \sqrt{\lambda + 1} \tag{2.16}$$

which leads to the so called square root safety stock formula. The order-up-to-level consists of the expected pipeline inventory $(\lambda + 1) \cdot \mu$ and the expected inventory just before the next order arrives, namely the safety stock $SS = k \cdot \sigma \cdot \sqrt{\lambda + 1}$. Therefore, the safety stock is given by the product of the service level driven safety factor k and the demand distribution convolution driven standard deviation of lead time demand.

Applying the standardization to the α-service level equation (2.13) yields $\Phi_{0,1}(k) = \alpha$ and $k = \Phi_{0,1}^{-1}(\alpha)$. The symmetry of the normal distribution ensures that the safety stock is non-negative as long as $\alpha \geq 0.5$. Since both the standard cumulative density $\Phi_{0,1}$ and the corresponding inverse $\Phi_{0,1}^{-1}$ cannot be solved analytically, numerical approximations are necessary to compute the required safety factor k. As a first approach, the equation $\Phi_{0,1}(k) = \alpha$ can be solved by bisection or by the Newton-Raphson method. For the evaluation of the cumulative density, the following approximation[52] can be used instead of numerical integration.

$$\Phi_{0,1}(k) = 1 - \phi_{0,1}(k)(b_1 w + b_2 w^2 + b_3 w^3 + b_4 w^w + b_5 w^5) + \epsilon(w) \quad (2.17)$$

where $|\epsilon(w)| < 0.000000075$, $w = 1/(1 + b_0 \cdot k)$ and the parameters b_i given in Table 2.3. In case of $x < 0$, the symmetry of the normal distribution yields $\Phi_{0,1}(x) = 1 - \Phi_{0,1}(-x)$.

Table 2.3. Parameters for the approximation of $\Phi_{0,1}$

i	b_{2i}	b_{2i+1}
0	0.231641900	0.319381530
1	-0.356563782	1.781477937
2	-1.821255978	1.330274429

Alternatively, k can be determined by directly approximating the inverse of the standard normal cumulative distribution function.[53]

$$k = w - \frac{a_0 + a_1 w + a_2 w^2}{1 + b_0 w + b_1 w^2 + b_2 w^3} + \epsilon(w) \quad (2.18)$$

where $|\epsilon(w)| < 0.00045$, $w = \sqrt{\ln(1/(1-\alpha)^2)}$, and the parameters a_i and b_i given in Table 2.4.

Table 2.4. Parameters for the standard normal inverse approximation

i	a_i	b_i
0	2.515517	1.432788
1	0.802853	0.189269
2	0.010328	0.001308

For a γ-service level constraint, the normal distribution standardization being applied to the expected backlog integral yields

$$\int_k^\infty (x - k)\phi_{0,1}(x)dx = \frac{(1-\gamma)\mu}{\sigma\sqrt{\lambda+1}}. \quad (2.19)$$

The integral in this expression is called the standardized loss function $G(k)$ and can alternatively be expressed in by

---

[52] HASTINGS, C. [1955], ABRAMOWITZ, M., STEGUN, I.A. [1970].
[53] HASTINGS, C. [1955], ABRAMOWITZ, M., STEGUN, I.A. [1970].

$$G(k) = \int_k^\infty (x - k)\phi_{0,1}(x)dx = \phi_{0,1}(k) - k(1 - \Phi_{0,1}(k)). \qquad (2.20)$$

In order to find the safety factor that satisfies

$$G(k) = \frac{(1 - \gamma)\mu}{\sigma\sqrt{\lambda + 1}},$$

bisection or Newton-Raphson iteration can be applied. For the numerical evaluation of $G(k)$, it is again necessary to use approximation (2.17) for the cumulative density function. If the unique[54] solution yields a negative safety factor, the application of Assumption 1 requires $k = 0$. For a normal distribution, a critical lead time τ can be derived analytically that indicates the maximum lead time that is covered by a zero safety stock without violating the service level constraint.

$$\tau = 2\pi \left((1 - \gamma)\frac{\mu}{\sigma}\right)^2 \qquad (2.21)$$

Instead of finding the required safety factor iteratively, the solution of equation (2.19) with respect to k and regarding Assumption 1 yields

$$k = \begin{cases} G^{-1}\left((1 - \gamma)\frac{\mu}{\sigma\sqrt{\lambda + 1}}\right) & \lambda + 1 > \tau \\ 0 & \lambda + 1 \leq \tau \end{cases}. \qquad (2.22)$$

For the inverse of the standardized loss function, SCHNEIDER[55] suggests the following approximation.

$$k(w) = \frac{a_0 + a_1 w + a_2 w^2 + a_3 w^3}{1 + b_0 w + b_1 w^2 + b_2 w^3 + b_3 w^4} + \epsilon(w) \qquad (2.23)$$

where $|\epsilon(w)| < 0.0002$, $w = \sqrt{\ln\left(\frac{5\sigma\sqrt{\lambda+1}}{\mu(1-\gamma)}\right)^2}$ and the parameters given in Table 2.5.

Table 2.5. Parameters for the approximation of the inverse of $G(k)$

i	a_i	b_i
0	-5.39255692478	-0.724964855323
1	5.62110547023	0.507326622469
2	-3.88368305104	0.0669136868273
3	1.08972995117	-0.00329129113899

In the β-service level case, the numerical analysis is complicated because the shortage expression is given by the difference of two integrals. Nevertheless, the order-up-to-level can be determined by bisection or Newton-Raphson

---

[54] It is easy to see that $dG(k)/dk < 0$.
[55] SCHNEIDER, H. [1979], p. 217.

iteration. For a given order-up-to-level S, both integrals have to be standard-ized differently with $k_1 = (S-(\lambda+1)\mu)/(\sigma\sqrt{\lambda+1})$ and $k_2 = (S-\lambda\mu)/(\sigma\sqrt{\lambda})$ to account for the one period time difference in the backlog integrals. This yields

$$\sigma \cdot \sqrt{\lambda+1} \cdot G(k_1) + \sigma \cdot \sqrt{\lambda} \cdot G(k_2) = (1-\beta)\mu.$$

Iterative numerical search algorithms are also applied when demands are assumed to be Mixed-Erlang. For the evaluation of the cumulative probability density and the loss function integral of a λ- or $(\lambda+1)$-fold convolution, two approaches can be followed alternatively. First, the exact distribution function of the convolution can be used which results in a mixture of several Erlang distributions. In order to reduce the problem to a mixture of two Erlang distributions, the approximation scheme discussed in 2.2.2.1 can be followed, that is a new mixture of two Erlang distributions is fitted on the expectation $(\lambda+1)\mu$ and the variance $(\lambda+1)\sigma^2$. This yields new parameters κ, q, and l. The equations for the determination of the order-up-to-level S that satisfies the considered service level constraint are the following:

- α-service level

$$ME_{\kappa,q,l} = \alpha \qquad (2.24)$$

- γ-service level

$$\int_S^\infty (x-S)\varphi_{\kappa,q,l}(x)dx = (1-\gamma)\mu \qquad (2.25)$$

The analytical solution of the loss integral is easily obtained to be

$$\int_S^\infty (x-S)\varphi_{\kappa,l,q}(x)dx = \frac{\kappa}{l}(1 - ME_{\kappa+1,l,q}(S)) - \frac{q}{\lambda}(1 - E_{\kappa,l})$$
$$-S(1 - ME_{\kappa,l,q}(S)).$$

- β-service level
The approximation procedure requires to fit a Mixed-Erlang distribution to both, the $(\lambda+1)$- and the λ-fold demand distribution. The respective distribution parameter sets are denoted by (κ_1, l_1, q_1) and (κ_2, l_2, q_2). The equation for order-up-to-level determination becomes

$$\int_S^\infty (x-S)\varphi_{\kappa_1,l_1,q_1}(x)dx - \int_S^\infty (x-S)\varphi_{\kappa_2,l_2,q_2}(x)dx = (1-\beta)\mu. \quad (2.26)$$

Both loss integrals can be solved analytically by the expression presented for the γ-service level constraint.

2.3.2 Batch Ordering

When products are replenished in batches, the transaction motive is combined with the safety motive of inventories. In comparison to lot-for-lot ordering, batch ordering decreases the probability of stockouts in the first periods of an order cycle and period based service level constraints are satisfied with less safety stock. For an optimal policy, the reorder point s and the order-up-to-level S (or implicitly the minimal lot-size $S - s$) have to be determined simultaneously. From a hierarchical planning point of view, the decision upon safety inventory is a strategic/tactical decision whereas lot-sizing is an operative decision. Therefore, the lot-size is predetermined by some static lot-sizing model and the determination of the safety stock at the higher decision level anticipates the given lot-size.

The traditional model for static lot-size determination is the economic order quantity (EOQ).[56] Given the expected demand rate μ, the fixed ordering costs K and the holding cost rate h per item and period, the average cost minimizing lot size EOQ is

$$S - s = EOQ = \sqrt{\frac{2\mu K}{h}}. \qquad (2.27)$$

Under a periodic review (s, S) control rule, the inventory position will in general be some amount below the reorder point when the next inventory review takes place. This phenomenon is called the undershot of the reorder point. Therefore, the quantity $S - s$ only represents the minimal order quantity that is placed only if the inventory position exactly equals s at the review point of time. Otherwise, the undershot is replenished in addition to the minimal lot size quantity and the average lot size $\overline{Q}$ is given by[57]

$$\overline{Q} = EOQ + \frac{\sigma^2 + \mu^2}{2\mu}. \qquad (2.28)$$

The economic order quantity neglects the interaction between the lot-size, lead time, and backordering. First, the batch size and the processing time to manufacture the batch may be related. Second, backorders are profitable under batching cost considerations to some extent because filling a cumulative backlog together with the next replenishment lot allows for a larger lot-size without increasing holding costs and therefore yields a larger fix cost degression. Taking into account these effects results in the multiplication of the EOQ replenishment quantity with some correcting factor.

- Finite production rate w
  The EOQ model assumes that the entire lot is available after the lead time and that this lead time is independent of the batch size. If the outcome of

---

[56] See e.g. LEE, H.L., NAHMIAS, S. [1993], p. 9-11.
[57] SCHNEIDER, H. [1981].

the production process and the batch size are connected, a distinction between open and closed production is necessary. Closed production implies
that items from the lot are available for customer demand or further processing only if the entire lot is finished. Under an open production framework, products become available continuously with a finite production rate
of w units per time period. The resulting optimal economic production
quantity (EPQ) is given by

$$EPQ = EOQ \cdot \sqrt{\frac{w}{w - \mu}} \qquad (2.29)$$

where the economic order quantity is multiplied by a correcting factor
larger than one.
- Backlogging cost p per item and time period
Under proportional backordering costs, the lot-size increases compared to
the economic ordering quantity and the timing of reordering is delayed
until some backlog has been cumulated. The resulting cost minimizing
replenishment quantity is

$$EQS = EOQ \cdot \sqrt{\frac{h + p}{p}}. \qquad (2.30)$$

A shortcoming in the application is that accounting difficulties may require
the backorder penalty cost rate to be replaced by a service level constraint.
A suggestion to overcome this problem is to replace the penalty cost rate
by the equivalent service level. As outlined in the lot-for-lot replenishment
situation under an α-service level constraint, the situation specific equivalence $\alpha = p/(p + h)$ holds.

$$EQS = \frac{EOQ}{\sqrt{\alpha}} \qquad (2.31)$$

Similar to the finite production rate situation, the economic order quantity
is multiplied by a correcting factor larger than one.

When inventory is replenished in batches, the expressions for on-hand stocks,
shortages, and backlog at the end of an arbitrary period depend on the renewal equations as stated in Section 2.2.5. In order to derive more tractable
expressions, approximations derived from an expansion $Q \to \infty$ are used to
determine the reorder level s.[58] The error induced when applying these approximations is rather small as long as the minimal replenishment quantity is
sufficiently large. SCHNEIDER [1981] recommends the following approximate
equations if $\overline{Q} > 1.5\mu$ holds. Otherwise, the replenishment process can be
approximated by a lot-for-lot control rule since the expected lot-size almost
equals the expected period demand.

---

[58] SCHNEIDER, H. [1981].

- α-service level

$$\int_s^\infty (x-s)f_{\lambda+1}(x)dx = (1-\alpha)\overline{Q} \qquad (2.32)$$

- γ-service level

$$\int_s^\infty (x-s)^2 f_{\lambda+1}(x)dx = 2\mu(1-\gamma)\overline{Q} \qquad (2.33)$$

- β-service level

$$\int_s^\infty (x-s)^2 f_{\lambda+1}(x)dx - \int_s^\infty (x-s)^2 f_\lambda(x)dx = 2\mu(1-\beta)\overline{Q} \qquad (2.34)$$

with $\overline{Q}$ from (2.28). The computational solution of the three equations is similar to lot-for-lot ordering. For an α-type constraint and normally distributed demands, the standardization yields the equation

$$G(k) = \frac{(1-\alpha)\cdot\overline{Q}}{\sigma\sqrt{\lambda+1}} \qquad (2.35)$$

which can be solved by the two proposed iterative methods or by applying the approximation for G^{-1}. Analogously to lot-for-lot ordering under γ-type service level constraints, the critical replenishment lead time that can be covered with zero safety stocks becomes

$$\tau = 2\pi \cdot \left((1-\alpha)\frac{\overline{Q}}{\sigma}\right)^2. \qquad (2.36)$$

Larger lot-sizes $\overline{Q}$ offer an increasing coverage potential and if a sufficiently large number of period demands is jointly replenished, no safety stock is required to satisfy the service level constraint.

For Mixed-Erlang demands and an α-service level constraint, equation (2.32) can be solved iteratively as outlined for lot-for-lot ordering under a γ-service level constraint.

The analysis of the γ-service level equation differs from the expressions analyzed so far because of the appearance of the quadratic loss integral. For normally distributed period demand, the standardized quadratic loss integral $J(k)$ is defined and solved by

$$J(k) = \int_k^\infty (x-k)^2 \phi_{0,1}(x)dx = (1+k^2)(1-\Phi_{0,1}(k)) - k\phi_{0,1}(k). \qquad (2.37)$$

Then, the safety factor that satisfies equation (2.33) is determined from

$$J(k) = (1-\gamma)\frac{2\mu}{\sigma^2(\lambda+1)}\overline{Q} \qquad (2.38)$$

which can be solved iteratively by bisection or the Newton-Raphson method. As a consequence from the non-negativity assumption of the safety stock, the critical lead time (plus review period) that is covered with a zero safety stock level results from $J(0) = 0.5$. Solving (2.38) for $(\lambda + 1)$ yields

$$\tau = \frac{4\mu(1-\gamma)}{\sigma^2}\overline{Q}. \tag{2.39}$$

The solution with respect to k by inverting $J(k)$ and considering the non-negativity constraint gives

$$k = \begin{cases} J^{-1}\left(\dfrac{(1-\gamma)2\mu\overline{Q}}{\sigma^2(\lambda+1)}\right) & \lambda+1 > \tau \\ 0 & \lambda+1 \leq \tau \end{cases} . \tag{2.40}$$

For the inverse of $J(k)$ the following approximation applies instead of using an iterative numerical method.[59]

$$k = J^{-1}(x) = \frac{a_0 + a_1 w + a_2 w^2 + a_3 w^3}{b_0 + b_1 w + b_2 w^2 + b_3 w^3} + \epsilon(w) \tag{2.41}$$

where $|\epsilon(w)| < 0.00023$ (within the interval $-4 \leq k \leq 4$). The parameters are given in Table 2.6.

Table 2.6. Parameters for the approximation of the inverse of $J(k)$

| | | $x < 0.5$ | | $x \geq 0.5$ | |
| | | $w = \sqrt{\ln(1/x^2)}$ | | $w = x$ | |
i	a_i	b_i	a_i	b_i
0	-0.4188413	1	1.125946	1
1	-0.2554696	0.2134080	-1.319002	2.836738
2	0.5189103	0.04439934	-1.809643	0.6559378
3	0	-0.002639787	-0.1165009	0.008220435

For Mixed-Erlang distributed demands, equation (2.33) has to be analyzed iteratively. The quadratic loss integral for a mixture of two Erlang distributions with parameters (κ, l, q) becomes

$$\int_s^\infty (x-s)^2 \varphi_{\kappa,l,q}(x)dx = \frac{\kappa^2}{l^2}(1 - ME_{\kappa+2,l,q}(s)) - 2s\frac{\kappa}{l}(1 - ME_{\kappa+1,l,q}(s))$$

$$+s^2(1 - ME_{\kappa,l,q}(s)) + \frac{(1-q)\kappa}{l^2}(1 - E_{\kappa+2,l}(s)) - \frac{q\kappa}{l^2}(1 - E_{\kappa+1,l}(s))$$

$$+2s\frac{q}{l}(1 - E_{\kappa,l}(s)).$$

A β-service level constraint can be dealt with in the same way as proposed for lot-for-lot ordering.

---

[59] SCHNEIDER, H. [1981].

2.3.3 Stochastic Lead Times and Capacity Constraints

An important extension concerns the situation when lead times are stochastic. Because the analytical derivation of the lead time demand distribution is rather difficult when demands and lead times are both stochastic, the lead time part of the uncertainty is incorporated by adjusting the mean and the standard deviation of the deterministic lead time models with respect to the additional uncertainty between order release and order arrival according to (2.5) and (2.6).

Another important factor for safety stock planning, especially in multi-item production/manufacturing systems, is set by capacity constraints. Limited storage space for inventory together with capacitated work stations for processing jobs and multiple products that compete for these resources add on the complexity of appropriate safety stock level determination.

A storage space constraint is reasonable only in a multi-product or a multi-level situation where different products compete for scarce storage space. For a single product, such a constraint to limit the maximum inventory level implies an upper bound for the order-up-to-level S. If an order-up-to-level or the maximum inventory level S under batch ordering exceeds the available capacity, adjustments are necessary to find a feasible solution. Nevertheless, under lot-for-lot replenishments and a service level constraint, no adjustments will yield a feasible solution since both restrictions are conflicting, that is the service level increases with higher safety stocks and the same holds for the violation of the capacity constraint. Within a penalty cost approach, a feasible solution with respect to the capacity constraint is generated by increasing backorders and therefore penalties. The same reasoning applies if products compete for a joint storage space capacity. Under batch replenishment policies, the storage capacity utilization can be influenced by the determined lot sizes. A capacity violation may be eliminated by a reduced order quantity. On the other hand, reducing the order quantity demands for larger safety stock requirements. In the multi-product batch ordering case, a priority rule for decreasing the single lot-sizes is required but the solution of this problem is beyond the scope of this work.

A second aspect of capacity constraints concerns processors that serve several products. The available workload of different jobs for different products that are waiting for a processor, the priority rule that is applied by the processor for selecting the next job to be processed, and the size and time requirements of the jobs highly influence the lead times. As suggested by KARMARKAR[60] this interrelation between products, processing, priorities, capacities, and lot-sizes on one side and the resulting lead times on the other side can be modeled by capacitated queuing systems. By the analysis of such a queuing model, expected value and standard deviation of the throughput time can be used as estimators for stochastic lead time characteristics and

---

[60] KARMARKAR, U.S. [1987], [1993].

safety stock requirements can be derived from the above mentioned stochastic
lead time considerations.

2.3.4 Lost Sales Inventory Model

Complete backordering of unsatisfied customer requests and delivery from the
next available incoming orders represents one extreme in demand management. The contrary customer behavior as opposed to waiting for the delivery
is the lost sales situation. If product substitutes exist or if the customer demand was inspired by a desire which requires an instantaneous satisfaction,
the insufficiency of inventory causes a lost sale. In comparison to backorder
inventory models, adjustments for safety stock determination are necessary
if unsatisfied demand is lost. Additional problem complexity arises if several
outstanding orders exist which is especially the case under lot-for-lot inventory policies. In these cases, the determination of the demand quantity that
is lost in an arbitrary period requires the knowledge of the exact timing of
demands and incoming orders within the replenishment lead time. Therefore,
the optimal inventory control rule under lost sales will, in general, not be of
a single or two critical number type[61] (as it is the case for backordering systems) and will, in the worst case, depend on all outstanding orders.[62] Though
the optimal policy will be more complex, standard order-up-to-S and (s, S)
control is applied in the following and the lost sales impact is incorporated
by adjustments within the policy parameter determination.

In order to incorporate lost sales into required safety stocks, approximations for the expected lost sales in an arbitrary period and a replenishment
cycle are required. One possibility is to use the expected shortage in an arbitrary period.

- Expected period lost sales for lot-for-lot replenishment

$$E[LS_1(S)] = \int_S^\infty (x - S) f_{\lambda+1}(x) dx - \int_S^\infty (x - S) f_\lambda(x) dx \qquad (2.42)$$

- Expected cycle lost sales for lot-for-lot replenishment

$$E[LS_\lambda(S)] = \int_S^\infty (x - S) f_\lambda(x) dx \qquad (2.43)$$

- Expected period lost sales for (s, S) replenishment

$$E[LS_1(s)] = \frac{1}{2\mu} \left(\int_s^\infty (x - s)^2 f_{\lambda+1}(x) dx - \int_s^\infty (x - s)^2 f_\lambda(x) dx \right) \qquad (2.44)$$

---

[61] NAHMIAS, S. [1979].
[62] WHITTEMORE, A.S., SAUNDERS, S.C. [1977].

- Expected cycle lost sales for (s, S) replenishment

$$E[LS_\lambda(s)] = \frac{1}{2\mu} \int_s^\infty (x - s)^2 f_\lambda(x) dx \qquad (2.45)$$

These expressions are approximations that overestimate the exact lost sales. Demands are already lost in the first λ periods and incoming items that are used for satisfying backorders in the backordering model now remain in stock and reduce the expected lost sales from $t + \lambda$ to $t + \lambda + 1$. Therefore, the adjustments should only be applied if the service level is sufficiently large.[63] Then, the approximation error will be small.

For α- and β-type service level constraints under lot-for-lot and batch ordering, the corresponding service level equations are given in the following. The γ-service level is not relevant in the lost sales case since accumulation of shortages does not occur. Under an α-service level constraint, the available inventory to satisfy the cumulative demand $D_{\lambda+1}$ consists of the order-up-to-level plus the expected lost sales over the first λ periods. Therefore,

$$Prob\left\{S - D_{\lambda+1} + \int_S^\infty (x - S)f_\lambda(x)dx \geq 0\right\} = \alpha. \qquad (2.46)$$

If inventory is replenished in batches, the expected cycle replenishment quantity $\overline{Q}$ is increased by the λ period lost sales.

$$\int_s^\infty (x - s)f_{\lambda+1}(x)dx = (1 - \alpha) \cdot \left(\overline{Q} + \frac{1}{2\mu} \int_s^\infty (x - s)^2 f_\lambda(x)dx\right) \qquad (2.47)$$

The incorporation of lost sales effects in the β-service level case yields a distinction of the denominator into average satisfied and average unsatisfied demand.[64] Therefore, the average shortage is divided by the sum of expected period demand and expected period shortage.

$$\beta = 1 - \frac{\int_S^\infty (x - S)f_{\lambda+1}(x)dx - \int_S^\infty (x - S)f_\lambda(x)dx}{\mu + \int_S^\infty (x - S)f_{\lambda+1}(x)dx - \int_S^\infty (x - S)f_\lambda(x)dx} \qquad (2.48)$$

This equation equivalently yields

$$\int_S^\infty (x - S)f_{\lambda+1}(x)dx - \int_S^\infty (x - S)f_\lambda(x)dx = \frac{1 - \beta}{\beta}\mu. \qquad (2.49)$$

The same reasoning and analysis when applied to batch replenishment situations yields the following equation for determining the reorder level.[65]

$$\int_s^\infty (x - s)^2 f_{\lambda+1}(x)dx - \int_s^\infty (x - s)^2 f_\lambda(x)dx = 2\mu\frac{1 - \beta}{\beta}\overline{Q}. \qquad (2.50)$$

---

[63] TIJMS, H.C. [1994], p. 61.
[64] TIJMS, H.C. [1994], p. 60.
[65] TIJMS, H.C. [1994], p. 66.

In both equations, the right hand side is larger than for the corresponding backorder situation. Therefore, the resulting safety stock to achieve the same service level will be smaller compared to the backorder case. Nevertheless, some care is necessary when using these relations since customer sensitivity will in general be higher in the lost sales case and therefore, it is rather unlikely that choosing the same service level size is appropriate for both cases.

The above equations can be solved by iterative numerical search along the same lines outlined for the backorder environment. SCHNEIDER[66] suggests a simple alternative for inventory systems under β-service level constraints. Order-up-to-levels are corrected by the expected lost sales.

- Lot-for-lot replenishment

$$S := S - (1 - \beta)\mu\lambda \tag{2.51}$$

- (s, S) replenishment

$$S := s + \overline{Q} - (1 - \beta)\overline{Q} = s + \beta \cdot \overline{Q} \tag{2.52}$$

In contrast to all other proposed methods, the order-quantity instead of the reorder point is adjusted.

2.3.5 Numerical Example

In the following, the previous safety stock determination equations are illustrated. A periodically reviewed stockpoint faces a demand process with expected value $\mu = 10$ units per period. The replenishment process follows lot-for-lot or batch ordering with an expected processing time of $\lambda = 2$ periods. The safety stock levels that are required in order to satisfy respective α-, γ-, or β-service level constraints are determined for different coefficients of demand variation v. For a given scenario of μ and σ, both, the normal and the Mixed-Erlang demand model are analyzed in order to illustrate the recommendations given when to apply the respective demand model.

The safety stock requirements increase with both, increasing demand variability and service level size. For normally distributed demands

$$SS_\alpha > SS_\gamma > SS_\beta$$

holds. In case of the Mixed-Erlang demand model, $SS_\gamma > SS_\beta$ holds whereas $SS_\alpha > SS_\gamma$ is only valid if demand variability is not too large.

The comparison of the normal and the Mixed-Erlang demand model stresses the negative demand problem of the normal distribution approximation. For small coefficients of variation, the safety stock difference between

---

[66] SCHNEIDER, H. [1981].

Table 2.7. Safety stock levels for lot-for-lot replenishment

	v	Normal					Mixed-Erlang				
		0.7	0.8	0.9	0.95	0.99	0.7	0.8	0.9	0.95	0.99
α	0.1	0.9	1.5	2.2	2.9	4.0	0.9	1.5	2.2	2.9	4.2
	0.2	1.8	2.9	4.4	5.7	8.1	1.7	2.9	4.5	5.9	8.6
	0.3	2.7	4.4	6.7	8.6	12.1	2.5	4.3	6.8	9.0	13.4
	0.4	3.6	5.8	8.9	11.4	16.1	3.2	5.6	9.2	12.2	18.4
	0.5	4.5	7.3	11.1	14.2	20.2	3.9	6.9	11.5	15.5	23.7
	0.6	5.5	8.8	13.3	17.1	24.2	4.5	8.2	13.9	18.9	29.2
	0.7	6.4	10.2	15.5	19.9	28.2	5.0	9.5	16.2	22.3	35.0
	0.8	7.3	11.7	17.8	22.8	32.2	5.5	10.7	18.6	25.8	41.0
	0.9	8.2	13.1	20.0	25.6	36.3	5.9	11.8	20.9	29.3	47.2
γ	0.1	0.0	0.0	0.0	0.4	2.1	0.0	0.0	0.0	0.4	2.1
	0.2	0.0	0.0	0.9	2.4	5.2	0.0	0.0	0.9	2.5	5.6
	0.3	0.0	0.2	2.7	4.8	8.7	0.0	0.1	2.9	5.2	9.8
	0.4	0.0	1.7	4.8	7.4	12.4	0.0	1.8	5.3	8.4	14.6
	0.5	1.0	3.5	7.1	10.3	16.3	1.0	3.8	8.1	12.0	20.0
	0.6	2.5	5.4	9.6	13.2	20.3	2.7	6.0	11.2	16.0	26.0
	0.7	4.3	7.4	12.2	16.3	24.4	4.7	8.6	14.7	20.4	32.6
	0.8	6.1	9.6	14.9	19.5	28.6	7.0	11.4	18.5	25.2	39.8
	0.9	8.1	11.9	17.7	22.8	32.8	9.4	14.5	22.6	30.4	47.5
β	0.1	0.0	0.0	0.0	0.4	2.1	0.0	0.0	0.0	0.4	2.1
	0.2	0.0	0.0	0.9	2.4	5.2	0.0	0.0	0.9	2.5	5.6
	0.3	0.0	0.1	2.7	4.8	8.7	0.0	0.1	2.8	5.2	9.8
	0.4	0.0	1.6	4.8	7.4	12.4	0.0	1.5	5.1	8.3	14.6
	0.5	0.5	3.2	7.0	10.2	16.3	0.0	3.1	7.7	11.7	19.9
	0.6	1.8	5.0	9.4	13.2	20.3	1.1	4.8	10.4	15.4	25.7
	0.7	3.2	6.8	11.9	16.2	24.4	2.3	6.7	13.3	19.4	32.0
	0.8	4.6	8.7	14.5	19.3	28.5	3.5	8.6	16.4	23.6	38.7
	0.9	6.2	10.7	17.1	22.5	32.7	4.9	10.7	19.8	28.2	46.0

the two models is rather small, whereas for $v > 0.4$ significant differences occur and the safety stocks determined under the normal demand assumption are too small.

The safety stock planning example under batch ordering assumes the same problem data except for a predetermined batch size of $\overline{Q} = 30$. This implies that three periods average demands are batched. The results in Table 2.8 show that, compared with lot-for-lot replenishment, safety stocks are required only for large service levels and large coefficients of demand variability and that a significant reduction in the required safety stock size is possible.

Complementary to the cases where the non-negativity assumption for safety stocks applies, the critical coverage horizons τ are determined. These values characterize the maximum replenishment lead time (including the review period) that can be covered without holding safety stocks. For normally distributed demands and an order-up-to-policy, the results are given in Table 2.9. The additional safety potential that is offered by cycle inventories under an (s, S)-policy generates positive coverage horizons also for the α-service

Table 2.8. Safety stock levels for batch replenishment

	v	Normal					Mixed-Erlang				
		0.7	0.8	0.9	0.95	0.99	0.7	0.8	0.9	0.95	0.99
α	0.1	0.0	0.0	0.0	0.0	1.0	0.0	0.0	0.0	0.0	1.0
	0.2	0.0	0.0	0.0	0.0	3.4	0.0	0.0	0.0	0.0	3.6
	0.3	0.0	0.0	0.0	1.3	6.2	0.0	0.0	0.0	1.3	6.8
	0.4	0.0	0.0	0.0	3.1	9.2	0.0	0.0	0.0	3.3	10.5
	0.5	0.0	0.0	1.0	5.1	12.4	0.0	0.0	1.0	5.6	14.7
	0.6	0.0	0.0	2.5	7.2	15.7	0.0	0.0	2.7	8.2	19.3
	0.7	0.0	0.0	4.3	9.5	19.1	0.0	0.0	4.7	11.2	24.4
	0.8	0.0	0.0	6.1	11.9	22.6	0.0	0.0	7.0	14.4	29.9
	0.9	0.0	0.4	8.1	14.4	26.2	0.0	0.2	9.4	17.9	36.0
γ	0.1	0.0	0.0	0.0	0.0	0.0	0.0	0.0	0.0	0.0	0.0
	0.2	0.0	0.0	0.0	0.0	0.0	0.0	0.0	0.0	0.0	0.0
	0.3	0.0	0.0	0.0	0.0	2.4	0.0	0.0	0.0	0.0	3.0
	0.4	0.0	0.0	0.0	0.0	5.3	0.0	0.0	0.0	0.0	6.6
	0.5	0.0	0.0	0.0	1.2	8.4	0.0	0.0	0.0	2.2	11.0
	0.6	0.0	0.0	0.0	3.6	11.8	0.0	0.0	0.5	5.6	16.1
	0.7	0.0	0.0	1.5	6.2	15.4	0.0	0.0	3.6	9.5	22.1
	0.8	0.0	0.0	3.9	9.0	19.1	0.0	0.0	7.2	14.0	28.8
	0.9	0.0	0.1	6.4	12.0	22.9	0.0	3.0	11.3	19.1	36.4
β	0.1	0.0	0.0	0.0	0.0	0.0	0.0	0.0	0.0	0.0	0.0
	0.2	0.0	0.0	0.0	0.0	0.0	0.0	0.0	0.0	0.0	0.1
	0.3	0.0	0.0	0.0	0.0	2.4	0.0	0.0	0.0	0.0	2.7
	0.4	0.0	0.0	0.0	0.0	5.3	0.0	0.0	0.0	0.0	6.5
	0.5	0.0	0.0	0.0	1.0	8.4	0.0	0.0	0.0	1.8	10.8
	0.6	0.0	0.0	0.0	3.4	11.7	0.0	0.0	0.0	4.8	15.7
	0.7	0.0	0.0	0.8	5.9	15.3	0.0	0.0	1.9	8.3	21.4
	0.8	0.0	0.0	3.0	8.6	19.0	0.0	0.0	4.9	12.2	27.7
	0.9	0.0	0.0	5.4	11.4	22.8	0.0	0.0	8.3	16.8	34.8

level case (Table 2.10) and yields larger coverage horizons for γ- and β-service levels compared to the order-up-to-policy (Table 2.11).

Table 2.9. τ-values for normally distributed demands and lot-for-lot orders

v	γ					β				
	0.7	0.8	0.9	0.95	0.99	0.7	0.8	0.9	0.95	0.99
0.1	56.6	25.1	6.3	1.6	0.1	81.2	26.3	6.3	1.6	-
0.2	14.1	6.3	1.6	0.4	0.0	19.4	6.5	1.6	-	-
0.3	6.3	2.8	0.7	0.2	0.0	8.1	2.8	-	-	-
0.4	3.5	1.6	0.4	0.1	0.0	4.2	1.6	-	-	-
0.5	2.3	1.0	0.3	0.1	0.0	2.5	1.0	-	-	-
0.6	1.6	0.7	0.2	0.0	0.0	1.6	-	-	-	-
0.7	1.2	0.5	0.1	0.0	0.0	1.2	-	-	-	-
0.8	0.9	0.4	0.1	0.0	0.0	-	-	-	-	-
0.9	0.7	0.3	0.1	0.0	0.0	-	-	-	-	-

Table 2.10. τ-values for normally distributed demands and batch orders

			α		
v	0.7	0.8	0.9	0.95	0.99
0.1	508.9	226.2	56.6	14.1	0.6
0.2	127.2	56.6	14.1	3.5	0.1
0.3	56.6	25.1	6.3	1.6	0.1
0.4	31.8	14.1	3.5	0.9	0.0
0.5	20.4	9.1	2.3	0.6	0.0
0.6	14.1	6.3	1.6	0.4	0.0
0.7	10.4	4.6	1.2	0.3	0.0
0.8	8.0	3.5	0.9	0.2	0.0
0.9	6.3	2.8	0.7	0.2	0.0

Table 2.11. τ-values for normally distributed demands and batch orders

			γ					β		
v	0.7	0.8	0.9	0.95	0.99	0.7	0.8	0.9	0.95	0.99
0.1	360	240	120	60	12	794.9	418.7	155.4	65.6	12.0
0.2	90	60	30	15	3	196.8	103.4	38.3	16.2	3.0
0.3	40	26.7	13.3	6.7	1.3	86.0	45.1	16.6	7.1	1.3
0.4	22.5	15.0	7.5	3.8	0.8	47.3	24.7	9.1	3.9	-
0.5	14.4	9.6	4.8	2.4	0.5	29.4	15.2	5.6	2.4	-
0.6	10.0	6.7	3.3	1.7	0.3	19.7	10.2	3.7	1.7	-
0.7	7.4	4.9	2.5	1.2	0.2	13.9	7.1	2.6	1.2	-
0.8	5.6	3.8	1.9	0.9	0.2	10.1	5.2	2.0	-	-
0.9	4.4	3.0	1.5	0.7	0.2	7.6	3.8	1.5	-	-

The last numerical illustration refers to the safety stock requirements under joint demand and lead time uncertainty. Since this case is transformed into a pure demand uncertainty model by adjusting the lead time demand standard deviation, Table 2.12 exhibits the adjusted coefficient of variation over the lead time that depends on the single period coefficient of demand variation and the coefficient of lead time variation.

Table 2.12. Coefficient of variation under joint demand and lead time uncertainty

v_D/v_λ	0.0	0.1	0.2	0.3	0.4	0.5	0.6	0.7	0.8	0.9
0.0	0.00	0.07	0.13	0.20	0.27	0.33	0.40	0.47	0.53	0.60
0.1	0.06	0.09	0.15	0.21	0.27	0.34	0.40	0.47	0.54	0.60
0.2	0.12	0.13	0.18	0.23	0.29	0.35	0.42	0.48	0.55	0.61
0.3	0.17	0.19	0.22	0.26	0.32	0.38	0.44	0.50	0.56	0.62
0.4	0.23	0.24	0.27	0.31	0.35	0.41	0.46	0.52	0.58	0.64
0.5	0.29	0.30	0.32	0.35	0.39	0.44	0.49	0.55	0.61	0.67
0.6	0.35	0.35	0.37	0.40	0.44	0.48	0.53	0.58	0.64	0.69
0.7	0.40	0.41	0.43	0.45	0.48	0.52	0.57	0.62	0.67	0.72
0.8	0.46	0.47	0.48	0.50	0.53	0.57	0.61	0.66	0.71	0.76
0.9	0.52	0.52	0.54	0.56	0.58	0.62	0.66	0.70	0.74	0.79

With increasing lead time variability, the total coefficient of variation

$$\frac{\sqrt{\mu^2 \cdot \sigma_\lambda^2 + (\lambda + 1) \cdot \sigma_D^2}}{\mu \cdot (\lambda + 1)}$$

shows a degressive increment as a result of the joint risk pooling effect between demand and lead time variability.

2.3.6 Aspects for Safety Stock Reduction

The previous models for safety stock planning were directed to the problem of replenishment timing and size in a given demand and single supply environment. In many practical applications, the resulting stocks are regarded as unsatisfactorily large and inventory reduction programs are started. Besides replacing inefficient rules of thumb that generate too large stocks in order to to achieve target performance objectives, several possibilities for safety stock reduction are known from the literature and will be summarized below. The difference between these and the previous models is to modify the replenishment strategy and to influence the problem components that were regarded as data for the previous considerations.

In the traditional square root safety stock formula for stochastic demand, the main influencing factors are the replenishment lead time and the standard deviation of forecast errors (as an estimator for unexplained demand variability). Forecasting accuracy can be improved by using more elaborate statistical or econometric techniques. The objective is to explain more variation by incorporating more impact factors and, therefore, to reduce the remaining noise which necessitates safety stock. Lead time reduction initiatives are often connected with business process reengineering and improved information technology. The same argument holds if, in addition, the lead time itself is modeled by a random variable. One possibility of explaining part of lead time variability is to analyze more complex queuing models in order to relate lead time to scheduling rules and lot-sizes.

An alternative approach for safety stock reduction is to apply a different inventory control strategy. Instead of replenishing stocks by using a single supply mode, an allocation and assignment of the total requirements to several situation specific modes yields an opportunity for cost reduction and/or customer service improvement. Instead of replenishing from a single supplier, the purchase volume can be divided upon multiple suppliers with different supply contract conditions with respect to cost, lead time, and delivery reliability. In a two supplier situation with different costs and lead times[67] the supplier with the lower cost and the longer lead time is utilized for the regular replenishment volume whereas the expensive mode is used to react to larger demands in a shorter time. This additional option for faster reaction reduces

---

[67] WHITTEMORE, A.S., SAUNDERS, S.C. [1977], JANSSEN, F.B.S.L.P. [1998].

(if the application is profitable in general) the effective lead time and the additional purchase costs for the expensive alternative are traded off with safety stock reduction gains. The extreme strategy of this research stream are emergency ordering models where the expensive supply mode can be used with a one period lag or even instantaneously. In case of suppliers with different lead time variability or reliability, the use of both suppliers reduces the overall variability and therefore, the safety stock requirements compared to the use of a single supply mode. The same effect is achieved by lot splitting[68] where safety stock reductions are balanced with larger purchase prices from loosing purchasing power by allocating the total requirements volume to several suppliers. A slight modification of these modeling approaches is inherent to single supplier, multiple supply mode models. The basic example is the choice of the transportation mode, for instance ship versus plane. This idea can be extended by introducing the option of expediting outstanding orders in emergency requirements situations, either by changing priorities or rescheduling within shop floor control. In contrast to multiple supplier models or simple lead time reduction initiatives, these techniques require an operating flexibility in order to enable the use of several modes. While all the approaches mentioned so far implicitly reduce safety stock levels, KIMBALL's inventory control concept[69] directly assigns different replenishment options to inventory status situations. Safety stock connected with the regular replenishment mode is meant to cover demand variability up to some maximum reasonable demand level. Extraordinary large requirements are dealt with operating flexibility which can imply expediting pipeline orders as well as using a second external supplier.

As a last category of options for safety stock reduction, the following substitution alternatives are available. Safety inventory implemented to deal with uncertainty can be replaced by several other buffer sources. The first alternative to inventories is excess capacity.[70] Larger throughput requirements can be fulfilled in a shorter lead time if more processing capacity is available and therefore, the overall lead time variability is reduced by reducing the waiting time components of the lead time. A second source of safety inventory reduction potential can be explored from larger lot-sizes. As pointed out for determining safety stock levels in batch replenished inventory systems, an increasing batch size increases cycle inventory that can simultaneously serve for buffering purposes and therefore substitutes safety inventory. This effect is in contrast to the JIT initiatives where zero inventories are proposed by implementing a lot-size of one.[71] In a stochastic demand environment, such a cycle stock reduction must go hand in hand with increasing safety stocks in order to achieve the service level targets. A last substitution possibility

---

[68] KELLE, P., SILVER, E.A. [1990].
[69] KIMBALL, G.E. [1988].
[70] SOUTH, J.B., HIXSON, R. [1988].
[71] NATARAJAN, R., GOYAL, S.K. [1994].

is offered when considering the product and component design. Instead of constructing products with different components which all of them require safety inventory, standardization and concentration on some components offers benefits from component commonality. Even the use of more expensive standard components can be attractive due to the inventory reductions from risk pooling of component requirements.[72]

2.4 Feedforward Versus Feedback Inventory Control

Most inventory models utilize feedforward control for the determination of replenishment policies and safety stock levels. Forecasts for demand and lead time are derived and the parameters of some inventory control rule like (S)- or (s, S)-policy are determined. The forecast errors serve as a measure of demand and lead time variability which necessitate safety stocks. In general, these models are analyzed under stationary conditions which yields a constant safety stock level. Drawbacks in a feedforward approach are that every source of uncertainty like demand, lead time, and yield has to be considered by individual data analysis and resulting forecasts. Sometimes, even correlation between these figures may be present. Most models additionally assume known mean and standard deviations though the data results from estimations. In order to account for this additional source of uncertainty, model adjustments concerned with additional safety stocks are required to yield the required system performance.

Opposed to these shortcomings of conventional feedforward inventory control, the feedback technique provides a unified treatment of all sources of variability. Additionally, more emphasis is put on short run service level control instead of following cost minimization under a long run service level constraint.[73] The mechanism determines the realized service level over some given feedback horizon. If the short run service performance lies out of some tolerance interval $\underline{SL} \leq SL \leq \overline{SL}$, the safety stock is adjusted. It is assumed that future variability will be identical compared to the past and that the adjusted safety stock that would have been sufficient in the past will provide a sufficient service in the future. If the realized service level is below $\underline{SL}$, the safety stock level is increased until it would have been sufficiently large for past demand and delivery lead time characteristics in order to provide the required service level. If the realized service level is larger than required, a safety stock reduction is desirable until the upper tolerance level is reached for the short run demand and lead time history. Therefore, adjustments initiated by the feedback generate dynamic instead of static safety stock levels.

The advantage of feedback compared to the conventional approach is the unified reaction to all kinds of uncertainties, parameter misspecifications, or

---

[72] EYNAN, A. [1996].

[73] SPICHER, K. [1975], [1976], KÄSSMANN, G., KÜHN, M., SCHNEEWEISS, CH. [1986].

the wrong demand model and even dynamic changes in the model environment by the inherent learning of the required safety stock level. From an application point of view, these techniques monitor how practitioners determine required safety stock levels by trial and error from past experience and performance deviation driven adjustments. Simulations of a conventionally controlled inventory system provide the insight that service level performance is highly unstable at the beginning and often a large amount of 1000 or even more simulation periods is required until the service performance measure remains stable. From a management accounting perspective, such an outcome is highly undesirable since the reporting period will not be that large that service level stability is achieved. Target deviations have to be reported for each cycle and these are difficult to explain by stating that in the long run of several reporting periods the right performance will be hopefully achieved. Nevertheless, the feedback approach also faces some drawbacks compared to conventional inventory control. Computational effort increases and more data has to be handled. This aspect is almost negligible under todays information technology capabilities. A more serious shortcoming is that the approach is only numerical and therefore does not provide any analytical insight into the problem. In addition, some starting value for the safety stock is required and cannot be determined by the feedback idea. In general, the mechanism will fail for seasonal demand patterns since the adjustments react too late with respect to changing requirements cycles. The algorithmic implementation of the feedback idea differs for lot-for-lot and for batch-ordering. Under periodic replenishments, the order-up-to-level is given by the safety stock plus the expected pipeline inventory ($\lambda\mu$). The system dynamics can be evaluated for period shortages or backlogs in order to compute the required service level over the feedback time interval. The required safety stock level can be found by iterative bisection until the lower or upper interval value is reached. Service levels over the feedback interval $t = i - n + 1, .., i$ with safety stock iteration j are computed from

- α-service level

$$\alpha_i^j = \frac{\sum_{t=i-n+1}^{i} 1_{[y_i^j \geq 0]}}{n}, \tag{2.53}$$

- β-service level

$$\beta_i^j = 1 - \frac{\sum_{t=i-n+1}^{i} SH_t^j}{\sum_{t=i-n+1}^{i} d_t}, \tag{2.54}$$

- γ-service level

$$\gamma_i^j = 1 - \frac{\sum_{t=i-n+1}^{i} BL_t^j}{\sum_{t=i-n+1}^{i} d_t}. \tag{2.55}$$

For batch ordering, the computations are divided into several replenishment cycles. It is assumed that the replenishment timing remains unadjusted though different safety stocks imply different reorder points and, therefore, a different replenishment timing in the (s, S)-rule. Here, the feedback interval consists of $m = 1, ..., M$ replenishment cycles with $t = 1, ..., n(m)$ time periods. The system dynamics are recomputed only with respect to y_t, SH_t, and BL_t. Then, the above service level computations apply with $n = \sum_{m=1}^{M} n(m)$.

3. Materials Coordination in Supply Chains

In general, the single echelon stockpoint considered in Chapter 2 represents only part of a system of interacting stockpoints in a supply chain. If the supply and the customer side of the stockpoint belong to the same system, additional benefits can be realized from coordination mechanisms that go beyond pure replenishment activities and customer order satisfaction as assumed in the previous chapter. In the first part of this chapter, different types of multi-echelon systems and supply chains are formally introduced. In addition, practical applications and modeling topics behind the respective system are characterized together with its design options. In a system with several external demand points, dependencies that result from substitutional or complementary requirements have to be taken into account by the analysis of multiple product demand time series. The second part reviews the three main information and materials flow concepts for coordination of stockpoints in a multi-echelon system and discusses the role of safety stocks in each concept. The focus is put on the impact of uncertainty instead of synchronizing lot-sizes which itself concerns a complex problem area.[1] Both problem areas are often decomposed into a deterministic dynamic multi-echelon lot-sizing and a stationary stochastic buffer sizing optimization problem.

3.1 Multi-Echelon Production/Inventory Systems

A production/inventory system in comparison to a pure inventory system additionally addresses a multi-echelon product structure and/or joint capacity utilization of different processes. Therefore, the representation of a multi-echelon system can either describe the processor view to characterize the items processed on the same machine or the product structure view that defines the required materials to process a product. The following focuses on the product view. For the system representation, the alternative expression "bill of material" is used.

---

[1] MUCKSTADT, J.A., ROUNDY, R.O. [1993].

3.1.1 Classification and Modeling Aspects

According to the inventory classification in Chapter 2, inventories can be divided into purchased parts, work-in-process, and finished goods. Therefore, every stockpoint is assigned to one of the sets E, P, A.

A set of first-stage (i.e. most upstream) stockpoints without predecessors

E set of final-stage (i.e. most downstream) stockpoints without successors

P set of intermediate-stage stockpoints

From a single stockpoint's point of view, all stockpoints that are situated on succeeding stages of the supply chain are called downstream, whereas previous installations are called upstream stockpoints. The set E includes all stockpoints that face external customer demands for finished products as well as for spare parts which are required for service operations of replacing defective components of products that are in use. These stockpoints share that they only supply external customers instead of other supply chain stockpoints. If a stockpoint supplies both, other stockpoints and external customers, the two different types of requirements can be separated into the total requirements that are satisfied by an internal stockpoint and the external spare parts requirements at an additional final-stage stockpoint.

All stockpoints in the set A represent the externally purchased materials and the corresponding processes characterize the purchasing and supply processes. These stockpoints have no predecessor stockpoint in the modeled system. For the following modeling aspects, the sourcing strategy is not considered explicitly. Therefore, regardless of the number of supply modes for the respective material, the underlying product is assigned to a single stockpoint. Supplier reliability and quality considerations are included into the supply lead time and internal service level description. The sets A and E characterize the interface of the system to its environment and are the result of the modeling decision which processes and partners to include in the considered supply chain.

All stockpoints that are not included in the sets A and E belong to the set P of work-in-process stockpoints. These stockpoints represent products that have been generated by physical (manufacturing) or geographical (transportation) transformation processes and that have to be further used for other transformation processes.

Similar to the single-echelon model, the following notation is introduced in order to describe each stockpoint's process and performance characteristics. The problem of estimating and modeling the demand processes is addressed in the next subsection.

λ_i	(expected) processing time for product i
σ_{λ_i}	standard deviation of processing time
ST_i	delivery time for finished products ($\forall i \in E$)
h_i	holding cost per unit and period for i
α_i	stockout occurrence related service level for i
γ_i	stockout size and duration oriented service level for i

The processing time λ_i represents the time required to manufacture or to transport a replenishment order released by stockpoint i, given that the required material is available at all supplying stockpoints. Waiting times as a result of inventory insufficiency are not included. Process scheduling induced waiting times that depend on the assignment of processes to capacities may be included. The variability is measured by the respective standard deviation σ_{λ_i}. In addition to the supply lead time delay, a final-stage stockpoint $i \in E$ faces an admissible delay ST_i between customer order release and the corresponding fulfillment requirement. This service time will in general depend on the competition in the respective product market segment. In a make-to-stock manufacturing situation with immediate demand satisfaction requirement, ST is zero. In a make-to-order production environment, it equals the maximum cumulative processing time over all production stages.

The performance of each stockpoint is measured with respect to physical stocks that are valued with inventory holding costs h_i per item and per period on one side and customer service levels. Service level constraints for final-stage stockpoints reflect the penalty for delivery insufficiencies as outlined in Chapter 2. For internal stockpoints, the type and size of the respective constraint represents the degree of available flexibility in order to buffer against delivery insufficiencies to downstream stockpoints. The type, as well as the size of the constraints, may differ among the products of the system. Often, customer requirements are controlled by a fill-rate based service measure because penalties apply for each unsatisfied request. Internal service level constraints are chosen from a stockout occasion type because the general shortage situation causes (approximately) magnitude independent penalties.[2]

Next, the delivery interactions between the stockpoints are modeled. An index $i = 1, ..., n$ is assigned to every stockpoint by a low level coding procedure. Thereby, the next stockpoint to be numbered is characterized by $i \in A$ or by the fact that all its predecessors are already numbered. As a result, the stockpoints of the sets A, P, and E are consecutively numbered and finally, each stockpoint has an index that is higher than the indices of all its predecessor (supplying) stockpoints. The predecessor and successor relations between the stockpoints are described by the following notation.

---

[2] An example for such a service level constraint structure within a two-level distribution system is e.g. SCHNEIDER, H., RINKS, D.B., KELLE, P. [1995].

$n(i)$	set of direct successors to stockpoint i
$v(i)$	set of direct predecessors to stockpoint i
$A(i)$	set of first-stage products that are required by i (identical to $\{i\}$ for $i \in A$)
$E(i)$	set of final-stage products that require product i (identical to $\{i\}$ for $i \in E$)
$N(i)$	set of all downstream stockpoints to i (including i)
$V(i)$	set of all upstream stockpoints to i (including i)
w	set of stockpoints, represents a path from a stockpoint to another downstream one
$w(i,j)$	set of stockpoints included in one path from i to j (including i and j)
$W(A,i)$	set of all paths that start at the first stage and terminate at i
$W(A,E)$	set of all paths that start at the first stage and terminate at the final stage
$a_{i,j}$	production coefficient, that is the required material of product i in order to produce one item of j

The properties of these sets depend on the type of the underlying supply chain network which is discussed in the following. It should be mentioned that the category to which a system belongs highly depends on the modeling detail of included processes and aspects. The more aggregate the investigated model, the simpler will be the type of system network and vice versa.

3.1.1.1 Serial Systems. A serial multi-echelon system with n stockpoints consists of a single first-stage stockpoint $A = \{1\}$ and a single final-stage stockpoint $E = \{n\}$. The remaining $n - 2$ products are assigned to intermediate stockpoints $P = \{2, ..., n-1\}$. Except for the final stage, each stockpoint has a single direct successor $n(i) = i + 1$ and except for the first stage, each stockpoint has a single direct predecessor $v(i) = i-1$. Then, the sets of all successors and all predecessors are given by $N(i) = \{i, ..., n\}$ and $V(i) = \{1, ..., i\}$ respectively. Because of the single external source, single destination property of a serial system, $A(i) = \{1\}$ and $E(i) = \{n\}$. Additionally, a single path between two arbitrary stockpoints $i < j$ with $w(i,j) = \{m = 1, ..., n | i \le m \le j\}$ exists and $W(A,i) = w(1,i)$, $W(A,E) = w(1,n)$ holds. For a serial system, the low level coding procedure faces no degrees of freedom and yields a unique numbering of the stockpoints.

Most application examples for this simple type of a multi-echelon system stem from process industry where a final product is manufactured by several consecutive chemical reaction processes. Nevertheless, this example is more academic since it belongs to the serial type of network only if some level of aggregation is applied. At each chemical reaction stage, some substances, fluids, or solvents are added or required, and if the replenishment of these inputs is modeled in detail, the type of network will no longer be of the serial but of the convergent type.

For designing a serial supply chain, one important aspect concerns the identification of stockpoints, i.e. the degree of process aggregation. If the analysis of a given system identifies that at some stockpoints no buffer stocks are required, the corresponding processes can be consolidated with the next stock holding installation. Thereby, the system is aggregated to a simpler one with the same materials coordination policy.

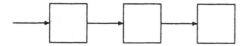

Fig. 3.1. Serial system

3.1.1.2 Divergent Systems. A divergent system contains a single supply input stockpoint ($A = \{1\}$) but several final-stage stockpoints ($|E| > 1$). Except for the first-stage, each stockpoint faces a single direct predecessor ($|v(i)| = 1$), but except for the final-stage ones, each stockpoint can have several direct successors ($|n(i)| \geq 1$). Caused by the non-uniqueness of successors, the low-level coding procedure contains some degree of freedom which reaches its maximum for the two-stage divergent system with a single first-stage and $n-1$ final-stage stockpoints. The minimum degree of freedom occurs in a $n-1$ stage adjusted serial system where one (arbitrary) stockpoint supplies an intermediate successor and a final-stage stockpoint or alternatively two final stockpoints. $A(i) = \{1\}$ reflects that each product results from the unique input stockpoint. The total number of paths through the network $|W(A, E)| = |E|$ is determined by the number of final stockpoints. The single direct predecessor property yields $|V(i)| = |w(1, i)|$.

Several applications of divergent systems can be found in physical distribution and manufacturing. In a single product distribution context, the divergent model focus starts with the central warehouse represented by the origin stockpoint. Depending on the number of distribution stages and the distribution strategy, products are shipped to distribution centers (sometimes in different countries) and from there to several regional warehouses, directly to regional warehouses, or even directly to the customers. In a manufacturing context, a divergent system occurs if several variants (e.g. different colors) are differentiated from the same base product. In this case, the initial stockpoint represents the split-off point. Different regional markets and customer classes with different service and delivery time priorities and requirements represent another field of application for divergent models.

The variety of application fields for divergent systems implies a number of design and reengineering topics. The developed model can serve as a building block for the evaluation of different alternatives concerning the materials coordination advantages and disadvantages of the respective system. The decision upon the distribution system design, i.e. the number of stages, number of warehouses at each stage, and the customer allocation to regional warehouses influences the inventory centralization/decentralization degree[3]. In a completely decentralized system, each regional warehouse operates independently from the others, that is requirements are individually replenished from an external supplier and separate buffers are held. Two main alternatives for reengineering the distribution system are stockpoint consolidation by centralizing replenishments and deliveries to customers to a smaller number of warehouses[4], and by implementing an additional central coordination stockpoint, where replenishments are coordinated by the central stockpoint and allocated to the other installations.[5] Thereby, an additional allocation problem of where and how much buffer inventory is required arises. In order to support these distribution system design decisions from the materials coordination perspective, appropriate divergent models are required.

Though the emphasis of warehouse location models is on trading-off capacity set-up cost with variable transportation cost, an inherent materials coordination aspect exists that is often neglected. The decision as to where to place warehouses and the corresponding allocation of customers to warehouses highly influences delivery lead times on the one hand and the demand characteristics with respect to level and variability on the other hand. This problem area also contains the problem of direct shipments of products to customers instead of utilizing the entire distribution chain which is discussed in the literature as the break quantity rule.[6] The aspect of redistribution and transshipments between stocking points refers to problems more related to multiple-mode supply considerations and is therefore not explicitly addressed in this work. Similar to subsection 2.3.6, these additional features in divergent systems can be regarded as strategies for safety stock reduction in multi-echelon systems.

In a manufacturing system design or reengineering initiative, divergent models provide assistance for the estimation of materials coordination costs induced by the product variety, for example, how much variants are distinguished for a particular base product and which benefits are drawn from reducing these to standard products or which additional operating costs are required to fulfill more differentiated customer needs. The latter topic is also discussed under the label of mass customization where almost individual products are manufactured by a system that can operate under mass production conditions. In order to enable such a strategy, the split-off point where the common base product is differentiated has to be determined. Therefore,

---

[3] See e.g. EVERS, P.T., BEIER, F.J. [1993], EVERS, P.T. [1995].
[4] EPPEN, G.D. [1979].
[5] EPPEN, G., SCHRAGE, L. [1981].
[6] KLEIJN, M.J., DEKKER, R. [1998].

alternative divergent systems have to be investigated for the evaluation of different postponement strategies.[7]

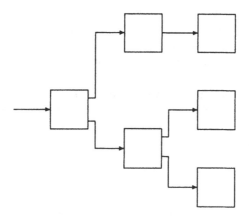

Fig. 3.2. Divergent System

3.1.1.3 Convergent Systems. A convergent multi-echelon system with n stockpoints has a single final-stage stockpoint $E = \{n\}$ but contains more than one first-stage stockpoints ($|A| > 1$). Except for the final-stage one, each stockpoint faces a single successor ($|n(i)| = 1$) but except for the first-stage ones, each stockpoint can have several predecessors ($|v(i)| \geq 1 \ \forall i \in P \cup E$). In contrast to the serial system, the low level coding procedure for numbering the stockpoints of a convergent systems contains some degrees of freedom. The two extreme cases with maximum and minimum degree of freedom for numbering the stockpoints are the two stage convergent system with a single final product being directly supplied by the other $n - 1$ stockpoints and an $n - 1$ stockpoint serial system with one (arbitrary) stockpoint being supplied by two instead of a single predecessor, respectively. All internal requirements are driven by the single final-stage stockpoint, therefore $E(i) = \{n\}$. The replenishments of each stockpoint i affect $|A(i)|$ first-stage supply activities and with respect to the single successor property, $|W(A, i)| = |A(i)| \geq 1$ holds. The final-stage requirements influence all stockpoint ($|W(A, E)| = |A|$). Since the destination path for each stockpoint is unique, the number of all successor stockpoints equals the number of points on the path from i to n ($|N(i)| = |w(i, n)|$).

Applications of convergent systems mostly stem from assembly operations where a single product is built from several components which themselves are

---

[7] LEE, H.L., TANG, C.S. [1997].

assembled from purchased parts. As already mentioned for serial system applications, if a chemical process is modeled in detail with all purchased materials that are added at each stage, this results in a convergent system. Design aspects in assembly systems mainly concern outsourcing decisions in general and the choice of suppliers and supply conditions. Outsourcing of manufacturing and assembly parts means to aggregate part of the convergent network to a single first-stage stockpoint with (in general) different process characteristics. Existing in- and outsourcing alternatives for components and available suppliers for required materials with different supply conditions concerning price and lead time (including lead time variability) can be evaluated with respect to their material coordination cost impact by comparing alternative convergent systems.

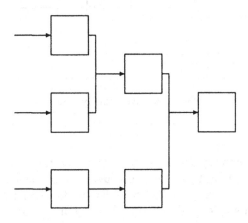

Fig. 3.3. Convergent system

3.1.1.4 General Acyclic Systems. The presented serial, convergent, and divergent systems represent the three basic types of multi-echelon systems where the serial is a special case of the convergent and the divergent one. Nevertheless, for modeling supply chain interaction in more detail and to cover several aspects that occur jointly, a general system is required. In this type of system, every stockpoint can have several predecessors ($|v(i)| \geq 1$) (except for purchased material) and successors ($|n(i)| \geq 1$) (except for final products). Therefore, a general system may contain several first-stage ($|A| \geq 1$) and final-stage stockpoints ($|E| \geq 1$). The low level coding procedure faces degrees of freedom as discussed for convergent and divergent systems. Especially the aspect of standardized components that are assembled to a product through different subassembly stages causes that there are multiple paths between two arbitrary stockpoints $|w(i,j)| \geq 1$. This aspect is not present in the three basic types of multi-echelon systems.

The modeling and analysis of several application topics requires a general network topology. If the supply chain consists of an assembly (convergent) manufacturing subsystem that ends in a central warehouse and a distribution (divergent) subsystem, this network is partly of basic types, but the system in total is of general type. Component commonality is a second application field already mentioned. Instead of using and coordinating the same material independently for all installations (where it is required) as it would be necessary within a convergent framework, the centralized coordination of common materials used for several assemblies causes that convergent and divergent characteristics occur jointly. Another aspect where divergence is introduced into mainly convergent manufacturing applications is the presence of independent demand (e.g. spare parts) for subassemblies. The characterization of independent external spare parts demand and internal demand posed by succeeding stockpoints is separated by introducing a second (final-stage) node that represents the independent requirements and therefore introduces divergence. In addition, multi-product manufacturing system modeling will require a general network.

For design and reengineering, especially for coordinating manufacturing and distribution operations or valuing the benefits of component commonality, a general network system provides a tool for evaluating the benefits and costs of different configurations and sourcing decisions.

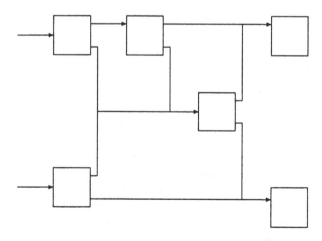

Fig. 3.4. General acyclic system

3.1.1.5 General Cyclic Systems. The previous non-cyclic system enables a low-level-coding of the stockpoints in an order where the next stockpoint to be numbered has only already numbered predecessors. This property no longer applies if cyclic dependencies exist, i.e. a predecessor arc of a stock-point originates from a direct or indirect successor of the stockpoint under consideration.

The practical modeling background behind cycles in multi-echelon systems are by- and co-products that are generated together with the desired product outcomes. Another application concerns returns of used products that are disassembled. The components are remanufactured for reuse as substitutes for regularly purchased parts at materials or subassembly stages. While the latter application mainly occurs in manufacturing and remanufacturing operations, by- and co- products especially are considered in process industries for chemical and pharmaceutical products. After some processing, e.g. cleaning or distillation of solvents, the by-products substitute input materials at other stockpoints. Depending on the reusing stockpoint, backward and forward reuse can be distinguished. Therefore, a cycle in the supply chain network occurs in case of backward reuse where materials are substituted at a direct or indirect predecessor of the process that generates the by-product.

In order to model and describe a network with cycles, some slight modifications of the general acyclic approach are necessary. First, the system without returns and by-products is described as a common forward acyclic network and the low level coding procedure applies. In a second step, the reuse arcs are inserted into the acyclic network. Therefore, a set R indicates all stockpoints where (by assumption) a single by-product occurs. Then, the set $bp(i)$ provides the information, at which stockpoints the by-product of process i is reused. The corresponding reuse quantity coefficient is given by a_{ji}^{bp}.

3.1.2 Demand Analysis

External and internal demands in a multi-echelon system with different products being manufactured at several levels provide new aspects in demand analysis. These are driven by the dependencies between the different product demands and by multi-stage processing. The independent analysis of the final-stage stockpoints concerning their demand characteristics can be performed along the lines presented for the single-echelon model, including customer behavior in shortage situations, determination of a satisfactory theoretical demand distribution, estimation of mean demand and the corresponding standard deviation, fitting of demand distribution parameters and the filtering of trend, seasonal, and special information effects. With respect to the demand interaction in a multi-item inventory system, a multiple item time series analysis is required. The most interesting key number within this context is the joint variability of arbitrary products that is measured by the covariance $\sigma_{i,j}$ or by the correlation coefficient $\rho_{i,j}$. A positive coefficient of correlation between two products indicates a complementary relationship, that is the requirement of one product (partly) induces a demand for the

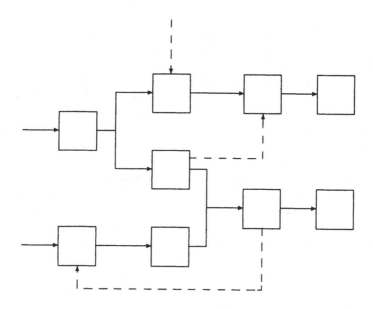

Fig. 3.5. General cyclic system

complement, whereas a negative coefficient of correlation indicates that one product (partly) serves as a substitute for the other product, and demand for one product implies (partly) lost requirements for the other product. In order to describe the external demand within a given theoretical demand model, the following notation is used.

μ_i	demand expectation for product i for an arbitrary (review) period
σ_i	standard deviation of demand for product i for an arbitrary (review) period
ρ_{ij}	correlation between demands for final products i and j in an arbitrary (review) period

From a given sample of past demand data, mean period demand and standard deviation can be estimated from (2.1) and (2.2) respectively. The required coefficients of correlation are estimated from the sample correlation coefficient.

$$\hat{\rho}_{i,j} = \frac{\sum_{t=1}^{\theta}(d_{i,t} - \mu_i)(d_{j,t} - \mu_j)}{\sqrt{\sum_{t=1}^{\theta}(d_{i,t} - \mu_i)^2 \cdot \sum_{t=1}^{\theta}(d_{j,t} - \mu_j)^2}} \qquad (3.1)$$

The presented multi-product demand model assumes that no correlation exists between demands of different time periods (for the same or even between different products) and that the external customer requests can be described independently from the service level and vice versa. The first aspect is analyzed in the demand forecast evolution approach of HEATH, JACKSON extended to multi-echelon divergent models by GÜLLÜ.[8]

The description of independent external demand serves as a basis for the analysis of internal requirements. Customer requirements cause products leaving the supply chain system and initiate internal replenishments. The relation between independent external and dependent internal demands is mainly influenced by the underlying concept of information and materials flow to be illustrated in more detail in Section 3.2. The information transmission of point-of-sale data observed by final-stage stockpoints is influenced by the applied inventory control rule which may introduce an additional time lag in data transmission if replenishment information is only transmitted to supplying stockpoints. In contrast, a direct point-of-sale data transmission to all affected stockpoints, regardless of delayed material flows, avoids additional uncertainty being introduced by the materials coordination system. The mean and standard deviation of demand of an arbitrary period under POS data aggregation, given the multi-stage production coefficients $a_{i,e}$, are represented by the following equations.

$$\mu_i = \sum_{e \in E(i)} a_{i,e} \cdot \mu_e \qquad (3.2)$$

$$\sigma_i = \sqrt{\sum_{e \in E(i)} \sum_{f \in E(i)} a_{i,e} \cdot a_{i,f} \cdot \sigma_e \cdot \sigma_f \cdot \rho_{i,j}} \qquad (3.3)$$

Under the assumption that aggregated demand follows the same or a chosen type of demand distribution (which may deviate from the assumed final-stage distributions[9]) a theoretical internal demand model can be estimated and developed from mean and standard deviation. Single-echelon considerations apply to internal demand analysis. Depending on the materials flow and operating in cases of non-sufficient material availability, all internal demand that exceeds available inventory can either be completely backlogged, completely lost in case of external emergency supply, or being modeled as a mixture of both.

3.2 Concepts of Information and Materials Flow

For a given model of a supply chain, the inventory control rule has to be specified. Several concepts are available to organize and to coordinate the underlying information and material flows. These alternative concepts can be classified into the categories of available local or central information upon

---

[8] HEATH, C.D., JACKSON, P.L. [1991], GÜLLÜ, R. [1996], [1997].
[9] SCHNEIDER, H., RINKS, D.B., KELLE, P. [1995].

stocks and requirements on the one hand and pull or push concepts for coordinating materials flow on the other hand. Further, program oriented concepts are distinguished from stochastic inventory control (SIC) approaches. The three most common concepts are (1) installation stock policies, a local information SIC pull approach, (2) echelon stock policies, a central information SIC approach and (3) material requirements planning, a central information production program oriented push concept. Alternative concepts that try to combine different features of these concepts are LRP and FiRST[10] and Cover-Time-Planning[11]

3.2.1 Installation Stock Policies

An installation stock policy is characterized by local operating of each stockpoint.[12] Available information contains only the locally available net inventory. Based on this, some single-echelon inventory control rules like (S) or (s, S) are applied to replenish stocks. The internal demand consists of the sum of all replenishment orders placed by directly supplied downstream stockpoints. In case of no lot-sizing, the installation stock policy represents Kanban ordering that is popular in the just-in-time philosophy. If available inventory does not enable the stockpoint to fulfill all requests, inventory is rationed among successors and missing parts are backordered.

Compared to the operation of a single installation, a locally controlled stockpoint that is part of a supply chain has to take into account delivery delays induced by material insufficiencies by at least one supplying installation. In a deterministic lead time, single-echelon model, it is assumed that the supplier can always deliver and therefore, the complete order arrives after the replenishment lead time. In a multi-echelon system, as a consequence of material shortages, the lead time becomes a random variable. This random variable consists of the processing time part, required to manufacture the product and given that all input materials are available, and a waiting time component for the time span between order release and complete availability of required materials.[13] In order to characterize the lead time demand distribution, the probability distribution of the waiting time or, alternatively, corresponding mean and standard deviation are required.

The role of safety stocks in an installation stock policy is (as in a single-echelon system) to protect individual installation service performance from demand variability. Demand for a first- or intermediate-stage stockpoint is determined from successor replenishments and these are not necessarily identical to the corresponding final-stage requirements. Reasons for replenishment

---

[10] VAN DONSELAAR, K. [1989].

[11] SEGERSTEDT, A. [1995].

[12] AXSÄTER, S. [1993] provides an overview on continuous review installation stock policies.

[13] TÜSHAUS, U., WAHL, C. [1998].

orders deviating from demands may include lot-sizing connected with replenishment of several period demands, increased order quantity due to expected capacity limitations in future periods, or purchase price or cost decreases induced by economies of scale. Therefore, replenishments in some periods increase, whereas, in others they decrease and, as a consequence, the overall demand variability measured by the period demand standard deviation increases. Then, more safety stocks are prescribed by the connection of buffer inventory to demand variability. This phenomenon that demand variability increases the more a stockpoint is situated upstream within the chain is called the bullwhip effect and is often observed in real world supply chains.[14] This disadvantage is based on the lack of real downstream point-of-sale data because of the local information and local inventory control. In addition safety stocks at every installation will be required. These facts are the main reasons of the general suboptimality of installation stock policies from the overall system point of view. The advantage of the application of installation stock policies is the easily implementable decentralized control of independent business units of a company or between several independent companies where a centralized control is hardly implementable. A second advantage is the decomposition of a complex system into single-echelon models which enable an easier finding of optimal or reasonable inventory control parameters. Nevertheless, the coupling of the decomposed units to the supply chain that is provided by the derivation of appropriate waiting time characteristics in general remains suboptimal.

3.2.2 Echelon Stock Policies

The shortcoming of using local information coupled with decentralized decision making in installation stock policies is avoided by echelon stock policies that operate under systemwide information. Different approaches within this concept are distinguished by the underlying type of information transmission and aggregation. The basic approach towards the echelon stock concept was presented by CLARK, SCARF.[15] The echelon stock of a stockpoint is defined as the systemwide inventory of the product assigned to that stockpoint, including the sum of physical stock at the stockpoint, all products being in transit to direct successors, and the echelon stock of all successors. For final-stage stockpoints, the echelon stock equals net inventory, that is physical stock minus backorders. The echelon stock provides aggregate information about direct or indirect influences on the stockpoint. Standard inventory control rules are applicable in the same manner as pointed out for a single echelon. An order is placed that raises the echelon inventory position to a desired order-up-to-level. The effect of decisions being based on echelon instead of on local information is that exactly the point-of-sale demands are replenished and no additional variability is introduced by individual replenishment

---

[14] FORRESTER, J.W. [1961], LEE, H.L., PADMANABHAN, V., WHANG, S. [1997].
[15] CLARK, A.J., SCARF, H. [1960].

decision making of all stockpoints that belong to the chain from the stock-point to the customer interface. The level to which systemwide inventory is increased with every replenishment is called the base-stock and as such the synonym base-stock policy is widely used for echelon stock policies. A shortcoming of the fact that each stockpoint operates under aggregate information becomes obvious if several successor subchains are present. If the available physical stock to supply all successors with the required amount of stock is not sufficient, rationing of the available inventory is necessary. The performance result of such a rationing mechanism with respect to service, inventory holding, and penalty costs depends on the individual inventory states of all successors. A rule that is only based on aggregate information will only provide an approximation. If, nevertheless, the rationing decision is based on aggregate information, the imbalance problem[16] can occur where a subchain is assigned with less echelon stock than it already received by past allocations.

The comparison of installation and echelon stock policies contains several categories. As an advantage, installation stock policies do not require an enlarged information basis which might be difficult to obtain due to the lack of information technology or organizational barriers. Additionally, a decentralized organization is supported. In contrast, echelon stock policies can be proven to be optimal for most basic multi-echelon systems that operate under a lot-for-lot replenishment regime without major setup costs. CLARK, SCARF proved this result for serial systems for a finite horizon and a discounted cost performance criterion.[17] The result remains valid under an infinite horizon as well as under an average cost criterion.[18] In a divergent system with several successors, the imbalance problem hampers the general optimality of this concept. If it is additionally assumed that the problem of negative shipment quantities does not occur (the so called balance assumption), the echelon stock policy is again optimal.[19] In a system of convergent type, optimality of the echelon stock coordination concept is obtained as a result of the equivalence of a convergent to a serial system.[20] DE KOK, VISSCHERS extend this equivalence result to more general networks with commonality where divergent and convergent dependencies occur jointly and prove the equivalence to divergent networks under the balance assumption.[21]

In supply chains with major setup costs at the stockpoints, the optimal materials coordination policy will in general be rather complex and depend on multiple inventory state variables as obtained by LUYTEN for two-stage

[16] ZIPKIN, P.H. [1984].

[17] CLARK, A.J., SCARF, H. [1960].

[18] FEDERGRUEN, A., ZIPKIN, P. [1984a].

[19] CLARK, A.J., SCARF,H. [1960], FEDERGRUEN, A., ZIPKIN, P. [1984b], LANGEN-HOFF, L.J.G., ZIJM, W.H.M. [1990].

[20] ROSLING, K. [1989], LANGENHOFF, L.J.G., ZIJM, W.H.M. [1990].

[21] DE KOK, A.G., VISSCHERS, J.W.C.H. [1999].

divergent systems.[22] Nevertheless, a simple (s, S) batch replenishment policy based on echelon stock information can be applied as an approximate extension to lot-for-lot environments.[23] When requirements are replenished in batches, another type of relation between installation and echelon stock policies is observed by AXSÄTER, ROSLING.[24] In case of stationary data without lot-sizing, both policies are equivalent. By an appropriate definition of installation stock inventory positions that include backorders for unsatisfied successor requests, the same materials flow as for an echelon stock policy is achieved. For any echelon stock policy with given parameters, parameters of an installation stock policy that implies the same materials flow exist and vice versa. This result no longer remains valid if lot-sizes are taken into account. Then, for serial and convergent systems, every installation stock policy can be replaced by an equivalent echelon stock policy, but the reverse does not hold. Therefore, the echelon concept is superior to installation stock policies in these cases. Nevertheless, for divergent systems this relation vanishes and examples for the superiority of both concepts can be found.[25]

The safety stock definition of expected net stock at a stockpoint just before the next order arrives can be alternatively formulated within the echelon stock concept. The aggregate echelon safety stock is defined as the stockpoint's safety stock plus all safety stock at downstream stockpoints. It provides a buffer resource to cover final-stage demand driven variability within the total subsystem, regardless of where the safety stock is exactly located within this subsystem. The general purpose of safety stocks within SIC concepts is to cover disturbances from uncertainty. This remains valid but is formulated in the same aggregate way as done for information and decisions.

A second concept that operates with immediate information transmission but assumes a different materials flow concept was suggested by KIMBALL.[26] The information flow follows the principle of demand explosion. Beginning with final-stage stockpoints, observed demand during the last review period is transmitted to all direct predecessors. Thereby, every stockpoint receives aggregate (echelon) information about the external demand of all products the stockpoint supplies directly or indirectly. Alternatively, external demands can be directly reported to all influenced upstream stockpoints. The difference between these two information flow alternatives is the degree of information aggregation. In the first variant, available information is only differentiated with respect to direct successors, whereas, in the second variant, demands for all corresponding final-stage stockpoints are individually available.

The materials flow induced by the information flow is characterized by the service time concept that replaces the application of an inventory con-

[22] LUYTEN, R. [1986].

[23] CLARK, A.J., SCARF, H. [1962].

[24] AXSÄTER, S., ROSLING, K. [1993], [1994].

[25] AXSÄTER, S. [1997], AXSÄTER, S., JUNTTI, L. [1996], [1997].

[26] KIMBALL, G.E. [1988]. See also MAGEE, J.F., BOODMAN, D.M. [1967].

trol rule. Every requirement reported to a stockpoint by its successors or the corresponding final-stage stockpoints is delivered after a deterministic, known service time. The interacting service times of all stockpoints represent a strategic decision to be taken together with system design. A special characteristic of the service time concept in contrast to CLARK, SCARF type echelon stock policies is the operation regardless to inventory availability. In situations of large requirements which in total exceed the available inventory, sufficient operating flexibility is assumed to be present in order to provide the stockpoint with the missing items and to enable the complete delivery of all requested material to successors.

The different materials flow implies a different role of safety stocks in the concept of KIMBALL. Since the service time concept is directed towards the strategic operation characteristics of a supply chain instead of resolving operative materials insufficiency problems, safety stocks are looked at from a strategic point of view. The main target of a strategic safety stock is to provide protection for reasonable variability. Adjustments in plans and operations should be avoided within the interval of normal variations. Within the CLARK, SCARF echelon stock concept, safety stocks are dimensioned with respect to all variations. In KIMBALL's concept, variations are divided into variations below a maximum reasonable level and extraordinary deviations. Strategic safety stocks are planned in order to cope with maximum reasonable deviations whereas extraordinary demands are dealt with operating flexibility and emergency reactions.

Though both approaches operate with an identical information background, the different assumptions concerning the materials flow imply advantages and disadvantages for the respective concept. The CLARK, SCARF approach can be regarded as the operative perspective concept where all material shortages immediately influence the materials flow to the succeeding stockpoints. Therefore, this concept does not require any operating flexibility or a distinction between reasonable and extraordinary variations which, in general, can only be made on a rather intuitive basis with managerial experience. On the other hand, the fact that every internal shortage results in backorders causes that material receiving stockpoints face a random lead time for their orders. Though the inventory control policy for the overall system is optimized, the individual manager at a stockpoint might feel to require more buffer in view of the variability of deliveries and may deviate from safety stocks prescribed from the echelon stock concept optimization model. This psychological effect which is another reason for the bullwhip effect and excess inventory counteracts the echelon stock philosophy. The service time concept avoids this shortcoming by prescribing a constant delivery time for all orders. It provides a better planning and understanding of the materials flow for each stockpoint. Instead of regarding and understanding the systemwide operation, only single-echelon inventory control operation is required. The disadvantage of this strategic approach is that the operative problems of dealing with extraordinary demand situations are not modeled explicitly.

3.2.3 Material Requirements Planning

In contrast to stochastic inventory control approaches, the Material Requirements Planning (MRP) concept[27] that refers to the work of ORLICKY[28] was developed for a deterministic make-to-order manufacturing system in order to provide a simple tool for the determination of replenishment quantities and timing. Uncertainty is not explicitly addressed because under a pure make-to-order manufacturing strategy, demand is given when material requirements are determined and other sources of uncertainty, including processing times and yield, are dealt with by replacing the random figures by certainty equivalent deterministic planning numbers. When the MRP concept is utilized in order to derive material requirements for make-to-stock driven manufacturing systems, the deterministic planning regime requires forecasts for uncertain demands.

The operation of an MRP system is based on the master production schedule. Given the demand forecasts for products, production quantities for all periods within the planning horizon are determined. Respective deviations of production from forecasts are justified by batching, production smoothing, and capacity constraints. The resulting production quantities determine the gross requirements. Available inventory to satisfy requests reduce these requirements whereas desired buffer stocks increase the required stock level. Orders that were released in the past are incorporated by subtracting these scheduled receipts from gross requirements. The resulting figure represents the net requirements to be replenished either lot-for-lot or, with respect to the presence of batching advantages, under some lot-sizing rule. The corresponding starting point for necessary operations results from the period where the products are required shifted by the planned processing time. In addition, these starting points determine the gross requirements period for all supplying stockpoints. From the product structure, supply relationships with adjoint material input coefficients can be observed. In an acyclic product structure network, stockpoints can be numbered by a low-level-coding procedure in a way that material requirements can be determined successively. Aggregate requirements for direct successor products determine gross requirements. The described steps net requirements calculation, lot determination, and process time shifting are repeated until all products have been dealt with. Thereby, infeasibilities may occur when materials would have been required in the past in order to guarantee the availability for some succeeding process. In these cases, additional considerations are required in order to obtain a

---

[27] For a brief introduction and literature overview of MRP components and aspects, see BAKER, K.R. [1993].

[28] ORLICKY, J. [1975].

feasible plan. The most simple activity is to use a present safety stock. Other activities like expedited processing or emergency ordering are not explicitly supported by the MRP framework and left to operational planning.

Deviations of random number realizations from planned figures in demands, lead times, and yield, are incorporated by the rolling horizon planning concept. Figures planned for the past period are replaced by realizations and planned inventory states are replaced by present ones. A new planning period is added and materials coordination is recalculated. Thereby, planned orders within the planning interval are adjusted depending on observed deviations. When requirements are replenished lot-for-lot, the adjustments concern order sizes. Batch replenishment may result in additional order releases or respective cancellations. Starting point for the calculation of requirements is the master production schedule (MPS) which prescribes production quantities for final products. Frequent replanning within the rolling horizon framework generates nervousness in the plans for final and previous stages.[29] Therefore, part of the plan is frozen and adjustments for the periods to be implemented next are prohibited.

Buffers in MRP systems[30] are explicitly or implicitly incorporated. The deterministic planning and calculation scheme necessitates planned lead times and input coefficients. If these are determined by the logic outlined in Chapter 2, where planned data contains the expected value increased by some safety surplus, replenishment orders will arrive earlier than planned and yield will be larger than expected, on an average basis. From the definition of safety stock, this will implicitly generate safety stocks. The same holds for buffers generated by hedging the MPS, where forecasted requirements are overestimated. Alternatively, safety stocks can be explicitly prescribed by planned safety stocks that are incorporated into the calculation as a surplus to gross requirements. The reasoning and justification for planned safety stocks differs for the stockpoints, depending on their location within the product structure network. At the final-stage customer interface level, safety stocks guarantee a required service level and enable the present MPS to be maintained. The frozen master schedule in order to avoid nervousness partly forbids replenishment adjustments at the final-stage product level. In situations of large positive deviations of demand realizations from their corresponding forecasts, demands can only be satisfied if excess requirements can be delivered from safety stocks. For the supply interface of purchased products, safety stocks are recommended for protection against delivery uncertainty and to guarantee purchased parts availability. Internal safety stocks provide a potential for the adjustment of replenishment quantities within the rescheduling and replanning module of a rolling horizon framework. The possibility of increasing lots for large demand occurrences requires some lot-sizing flexibility. If

---

[29] JENSEN, T. [1996].

[30] NEW, C. [1975], ETIENNE, E.C. [1987].

replenishments are made according to fixed ordering quantities, adjustments are not possible and internal safety stocks are dead stock and useless.

The question which buffer concept is appropriate for an MRP system depends on the sources of uncertainty. Comparisons between planned safety stocks and planned safety lead times are presented analytically by BUZACOTT, SHANTHIKUMAR and, within a simulation study, by WHYBARK, WILLIAMS.[31] The general result from these studies is to deal with time related uncertainty by safety times and to deal with quantity uncertainty by safety stocks. Another recommendation for safety time based buffers is made for systems with extreme lumpiness in demand or with large lot-sizes. Especially in these situations, the dead safety stock problem arises because most periods face zero demands and, therefore, the stocks are not required.

Though MRP differs from SIC concepts by the dominance of the deterministic framework, the concept can be compared to installation and echelon stock policies within the flexibility and replication framework. MRP-policies are regarded as (s, S)-policies with aggregated (echelon) stock information and policy parameters.[32] For serial and convergent systems, AXSÄTER, ROSLING[33] show that every echelon stock policy can be replicated by an MRP policy, but not vice versa. When the MRP concept is compared to KIMBALL's base stock concept, some additional similarities arise. The MRP concept does not provide any assistance in case of material shortages. Therefore, the strategic buffer motive that dominates within KIMBALL's concept to cope with reasonable deviations and to leave excess demand problems to be dealt with operating flexibility is shared by the MRP concept. In addition, the requirement for fixed replenishment lead times in order to calculate net requirements is shared by the constant service time assumption.

---

[31] BUZACOTT, J.A., SHANTHIKUMAR, J.G. [1994], WHYBARK, D.C., WILLIAMS, J.G. [1976].

[32] LAGODIMOS, A.G. [1990], LAMBRECHT, M.R., MUCKSTADT, J.A., LUYTEN, R. [1984].

[33] AXSÄTER, S., ROSLING, K. [1994].

4. Safety Stocks in Multi-Echelon Systems

All materials coordination concepts presented in 3.2 require buffers to cope with uncertainty. In this chapter, approaches for the determination of planned safety stocks for multi-echelon supply chains are discussed, developed, and extended. The additional complexity in multi-echelon networks compared to single-echelon inventory models results from two main sources. The first one is that due to large internal order releases, shortages at the supply stockpoint can occur that increase the replenishment lead time and therefore, require additional buffers. The second aspect is the possibility of allocating safety stocks over the entire system. For didactical purposes, the simplest version of a multi-echelon network, the serial system, is presented first, though it is only a special case of convergent and divergent networks.

4.1 Serial Systems

According to the presented variants of echelon stock materials coordination concepts, the respective safety stock planning procedures for basic types of serial systems are outlined in this section. In contrast to other approaches where buffers are set by some rules of thumb or by pure delivery service criteria, these methods derive safety stocks from a cost optimization and, therefore, an economic perspective. The full delay approach refers to the CLARK, SCARF[1] type of model where every shortage directly generates a delay in the materials flow. A safety stock planning procedure for a serial supply chain of arbitrary length was proposed by VAN HOUTUM, ZIJM[2]. No delay approaches refer to the coordination concept of KIMBALL[3] where material unavailabilities are covered by operating flexibility and do not generate an additional delay to the proposed service time. The determination of cost optimal safety stock norms for this concept is presented by SIMPSON[4].

[1] CLARK, A.J., SCARF, H. [1960].
[2] VAN HOUTUM, G.J., ZIJM, W.H.M. [1991].
[3] KIMBALL, G.E. [1988].
[4] SIMPSON, K.F. [1958].

4.1.1 Full Delay Approaches

4.1.1.1 The Model of Clark and Scarf. With respect to the data required for the CLARK, SCARF model, the following assumptions are made.

(1) The network is of serial type. Numbering of stockpoints is according to the notation introduced in 3.1.

(2) External customer demand can be modeled by any continuous demand distribution with density f and cumulative density F. Unsatisfied customer demands are backordered.

(3) Processing times for each stockpoint are described by integer constants. The final-stage processing time includes the review period.

(4) Material input coefficients are, without loss of generality, set equal to one for all stockpoints.

(5) The cost structure contains linear inventory holding and backordering penalty cost parameters per item and period. Setup costs are permitted only for the most upstream stockpoint.

(6) The external supplier can always deliver any requested amount of material. Therefore, no additional delay to the purchasing processing time occurs.

(7) Start-up effects are excluded from the analysis, i.e. the initial state of on-hand and pipeline stocks lies within the ranges generated by the optimal policy. This implies that the initial situation is a steady state and no overstocked installations exist.

In order to describe the optimal policy, CLARK, SCARF introduce the echelon stock concept. The echelon stock of a stockpoint is defined as the stock on hand at this stockpoint plus all stock in the downstream part of the serial system minus the final product's backlog. Within a cost optimization framework, it is shown that the optimal inventory control rule for all stockpoints is an echelon order-up-to policy, i.e. the echelon inventory position defined as the echelon stock plus the stock in process (from the preceding installation) is increased to a base-stock-level. In case of a positive setup cost for the most upstream stockpoint, this stockpoint operates under an echelon (s, S) control rule.

In order to characterize the system dynamics, single-echelon notation of state and decision variables is extended by introducing a stockpoint index that is assigned by a low-level coding procedure. The external supplier is denoted by $i = 0$. The state of the system is described at the end of an arbitrary period t after demand and before the next shipments arrive at the stockpoints.

$y_{i,t}$ net stock at stockpoint i at the end of period t
$OH_{i,t}$ on-hand stock at stockpoint i at the end of period t
$BL_{n,t}$ backlog of customer demand at the final-stage at the end of period t

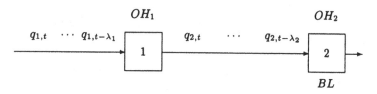

Fig. 4.1. State of a two-echelon serial system at the end of period t

Decisions of each stockpoint are made at the beginning of a period after the order placed λ_i periods ago has arrived. They concern the desired order and shipment quantities. In cases where available on-hand stock at the preceding stockpoint is insufficient to meet the desired order quantity, order and shipment sizes differ.

$x_{i,t}$ order placed by stockpoint i at the beginning of period t

$q_{i,t}$ shipment to stockpoint i initiated by $i-1$ at the beginning of t

The variables $q_{i,t-\lambda_i}, ..., q_{i,t-1}$ describe the state of the pipeline between stockpoints $i-1$ and i at the end of period $t-1$ just before the shipments $q_{i,t-\lambda_i}$ arrive. Echelon stock $\hat{y}_{i,t}$ and echelon inventory position $\hat{y}^p_{i,t}$ at the end of period t are given by (4.1) and (4.3) respectively. For the final stage, local and echelon stock definition coincide ($\hat{y}_{n,t} = y_{n,t}$).

$$\hat{y}_{i,t} = \sum_{j=i}^{n} OH_{j,t} + \sum_{j=i+1}^{n} \sum_{k=0}^{\lambda_j-1} q_{j,t-k} - BL_{n,t} \tag{4.1}$$

$$= OH_{i,t} + \sum_{j=0}^{\lambda_{i+1}-1} q_{i+1,t-j} + \hat{y}_{i+1,t} \tag{4.2}$$

Equation (4.2) represents an equivalent recursive definition of echelon stock. The echelon inventory position is defined by the amount of stock that becomes available during the next λ_i periods.

$$\hat{y}^p_{i,t} = \hat{y}_{i,t} + \sum_{j=0}^{\lambda_i-1} q_{i,t-j}. \tag{4.3}$$

Note that the echelon inventory position at the end of $t-1$ does not change until the ordering decision at the beginning of t is made.

The desired order quantities $x_{i,t}$ to increase the echelon inventory position of stockpoint i to the echelon order-up-to-level S_i are

$$x_{i,t} = (S_i - \hat{y}^p_{i,t-1})^+. \tag{4.4}$$

The maximum operator is not necessary if assumption (7) holds, that is if the initial echelon inventory position does not exceed the order-up-to-level.

The materials flow is restricted by available stock on hand at the preceding installation.

$$q_{i,t} = \min\{OH_{i-1,t-1} + q_{i,t-\lambda_i}; x_{i,t}\}, \ i = 2, ..., n \qquad (4.5)$$

$q_{1,t} = x_{1,t}$ represents the (uncapacitated) shipments of the external supplier.

The system dynamics (state transitions) from period t to $t+1$ are described by the following equations.

$$OH_{i,t} = OH_{i,t-1} + q_{i,t-\lambda_i} - q_{i+1,t} \qquad i = 1, ..., n-1$$
$$\hat{y}_{i,t} = \hat{y}_{i,t-1} - d_t + q_{i,t-\lambda_i} \qquad i = 1, ..., n$$
$$\hat{y}^p_{i,t} = \hat{y}^p_{i,t-1} - d_t + q_{i,t} \qquad i = 1, ..., n.$$

For stockpoint $i = n$, on-hand stock and backlog dynamics are given by

$$OH_{n,t} = (OH_{n,t-1} + q_{n,t-\lambda_n} - d_t)^+$$
$$BL_{n,t} = (BL_{n,t-1} + d_t - q_{n,t-\lambda_n})^+.$$

The following 8 period example illustrates the materials flow in a two-echelon serial inventory system that operates under the assumptions of CLARK, SCARF. The processing times are set to $\lambda_i = 2$ for both installations. The echelon order-up-to-levels are $S_1 = 60$ and $S_2 = 35$. For an expected period demand of $\mu = 10$ this implies identical local safety stock norms of $SS_i = 5$.[5] The initial state of the system at the beginning of period 1 is chosen to be equal to the expected steady state, i.e. shipment quantities $q_{i,t-j}$ are identical to the demand expectation $\mu = 10$ and on-hand stocks equal the planned safety stock norms. For a given demand scenario $(d_t)_{t=1,...,8}$, ordering and shipment decisions and on-hand stocks are listed in Table 4.1.

Internal inventory insufficiencies (marked with an asterisk where the desired order quantity $x_{2,t}$ deviates from the shipment quantity $q_{2,t}$) occur in periods 3 and 7 where 10 instead of 15, and 5 instead of 10 items can be shipped to stage 2. The 5 missing items in both cases are delivered together with the shipments in the 4th and 8th period, respectively. The realized customer side performance in this example, measured by empirical end item service levels, is $\hat{\alpha}=87.5\%$ and $\hat{\beta} = \hat{\gamma} = 93.75\%$.

The derivation of optimal stationary inventory control rule parameters requires for a construction of an average cost function.[6] A penalty cost p is incurred per item and period of a delivery delay at the final installation whereas no direct penalty is incurred for internal stock insufficiencies. Every stockpoint i assigns holding costs h_i to every item on hand at the end of a

---

[5] The figures serve to illustrate the materials flow. In an optimal policy for the CLARK, SCARF model with a finite planning horizon, the optimal order-up-to-levels will vary in time.

[6] For the following analysis see LANGENHOFF, L.J.G., ZIJM, W.H.M. [1990].

Table 4.1. Materials coordination example for the CLARK, SCARF model

		Stage 1				Stage 2			
t	d_t	$q_{1,t}$	$q_{1,t-1}$	$OH_{1,t}$	$x_{1,t+1}$	$q_{2,t}$	$q_{2,t-1}$	$y_{2,t}$	$x_{2,t+1}$
0	-	10	10	5	10	10	10	5	10
1	15	10	10	5	15	10	10	0	15
2	15	15	10	0	15	15	10	-5	15 (*)
3	5	15	15	0	5	10	15	0	10
4	5	5	15	5	5	10	10	10	5
5	20	5	5	15	20	5	10	0	20
6	10	20	5	0	10	20	5	0	10 (*)
7	5	10	20	0	5	5	20	0	10
8	5	5	10	10	5	10	5	15	5

period. Within the echelon stock concept, every stockpoint assigns holding costs to the echelon stock. Therefore, only the value added at stockpoint i is subject to inventory holding cost in order to avoid double counts. Let h_i^e denote the echelon holding cost parameter. Then, the traditional local inventory holding cost parameter is $h_i = \sum_{j=1}^{i} h_j^e$. The total costs for a given state at the end of an arbitrary period depend on whether there is a backlog or physical stock at the final stage. If no customer demands have been backordered ($y_n \geq 0$), holding costs $\sum_{i=1}^{n} h_i^e \cdot \hat{y}_i$ are assigned to positive echelon stocks. When the final stage faces a backlog ($y_n < 0$), penalty costs $-p \cdot y_n$ are added. Additionally, the backlog is included in echelon stocks $\hat{y}_i$ of all upstream installations. Therefore, holding costs calculated by a value added echelon stock accounting scheme underestimate the true holding costs by $\sum_{i=1}^{n} h_i^e y_n = h_n y_n$. Summarizing, the echelon cost function under a backlog is given by $\sum_{i=1}^{n} h_i^e \hat{y}_i - (p + h_n) y_n$.

Because the stock at internal and final stockpoints is a random variable due to random customer demand, the expected single period costs L_i that depend on the echelon stock $\tilde{y}_i$ at the beginning of a period are given by

$$L_n(\tilde{y}_n) = h_n^e \cdot \int_0^\infty (\tilde{y}_n - u) dF(u) + (p + h_n) \cdot \int_{\tilde{y}_n}^\infty (u - \tilde{y}_n) dF(u)$$

for the final stage $i = n$. The stockpoints $i = 1, ..., n - 1$ only assign holding costs.

$$L_i(\tilde{y}_i) = h_i^e \cdot \int_0^\infty (\tilde{y}_i - u) dF(u)$$

The overall multi-stage cost function is derived within a Dynamic Programming framework. Dynamic Programming value functions are defined recursively starting with the average cost function of the most downstream stage. Let $D_i(S_i, ..., S_n)$ denote the sum of average costs of installations i to n for given echelon order-up-to-levels S_i and assume that the echelon inventory position of installation i can always be increased to S_i, that is the supplier of stockpoint i can always deliver any requested amount of stock. For stage $i = n$, the echelon stock at the beginning of a period results from the order-up-to-level S_n by subtracting the demand within λ_n processing time periods.

$$D_n(S_n) = \int_0^\infty L_n(S_n - u)dF_{\lambda_n}(u)$$

Average costs of intermediate stockpoints are given by the direct average costs assigned to this stockpoint plus average downstream costs. Downstream costs are given by costs derived under the assumption that installation i can always deliver plus an additional penalty cost term for cases where this assumption is violated by internal delays. The additional penalty is expressed by the cost difference for cases where demand within λ_i periods exceeds the difference of the order-up-to-levels of stockpoints i and $i + 1$. Because the supplier of i can always deliver, the echelon inventory position of i can always be increased to S_i. Therefore, this stockpoint always replenishes the observed customer demand of the last period and the available echelon stock to increase the inventory position of stockpoint $i + 1$ equals the order-up-to-level S_i minus the material in the pipeline to i (which is identical to customer demand of the last λ_i periods). An internal delay occurs in situations where the available amount $S_i - D(\lambda_i)$ is smaller than the required amount S_{i+1}.

$$D_i(S_i, ..., S_n) = \int_0^\infty L_i(S_i - u)dF_{\lambda_i}(u) + D_{i+1}(S_{i+1}, ..., S_n)$$

$$+ \int_{S_i - S_{i+1}}^\infty [D_{i+1}(S_i - u, S_{i+1}, ..., S_n) - D_{i+1}(S_{i+1}, S_{i+2}, ..., S_n)]dF_{\lambda_i}(u).$$

The analysis of these Dynamic Programming functional equations results in the optimality of order-up-to policies because of the convexity of the value functions. The parameters S_i^* denote the base stock levels that minimize the equations. Due to the sequential structure of D_i, the values S_i^* can be determined sequentially beginning with the order-up-to-level of the most downstream stockpoint.[7] Therefore, n optimization problems that depend on one variable, instead of a complex problem with n variables, have to be solved resulting in a considerable reduction of solution complexity.

In general, the optimal policy is characterized by decreasing echelon order-up-to-levels

$$S_1^* \geq S_2^* \geq ... \geq S_n^*. \tag{4.6}$$

If $S_i < S_{i+1}$, the ordering and shipment policy implies that stockpoint i operates as a stockless inventory point, that is that every item on hand is immediately forwarded to the next downstream installation. Without affecting the materials flow of the policy, S_{i+1} can be lowered to the value of S_i. The repetition of this argument for all preceding stockpoints reveals that the consideration of adjusted order-up-to-levels defined by

---

[7] CLARK, A.J., SCARF, H. [1960]. This result is called the decomposition result.

$$\tilde{S}_{i+1} := \min\{S_1, S_2, ..., S_{i+1}\} \qquad i = 0, ..., n-1$$

is sufficient. This policy implies the same materials flow and is, therefore, associated with the same costs.

4.1.1.2 The Algorithm of Van Houtum and Zijm. VAN HOUTUM, ZIJM utilize the average cost function Dynamic Programming framework with its decomposition result to determine cost optimal policy parameters sequentially, beginning with the most downstream installation.[8] Though the algorithm presumes a penalty cost approach it can alternatively be applied in the service level case as described in 4.1.1.3.

The inability to increase the inventory position of the successor to the desired level results in additional penalty costs. The missing amount of stock for the shipment to stockpoint i is denoted by the shortfall random variable Y_i. Define the maximum physical stock level Δ_i of a non-final-stage stockpoint i by the difference of the echelon order-up-to-level of i and $i + 1$,

$$\Delta_i = \tilde{S}_i - \tilde{S}_{i+1}, \qquad \Delta_i \geq 0, \qquad i = 1, ..., n-1.$$

The shortfall of the most upstream stockpoint is zero by definition because the external supplier can always deliver. The shortfall of the second installation occurs when customer demand $D(\lambda_1)$ during the processing time of the first installation exceeds the maximum stock at the first installation,

$$Y_2 = (D(\lambda_1) - \Delta_1)^+.$$

The shortfall of other stockpoints can be described recursively. The echelon inventory position of i can be increased to the level $\tilde{S}_i - Y_i$. Therefore, the available amount of echelon stock is given by $\tilde{S}_i - Y_i - D(\lambda_i)$ and the recursive relation for the shortfall becomes

$$Y_{i+1} = (Y_i + D(\lambda_i) - \Delta_i)^+.$$

Successive substitution yields

$$Y_{i+1} = (...(D(\lambda_1) - \Delta_1)^+ + D(\lambda_2) - \Delta_2)^+... + D(\lambda_i) - \Delta_i)^+.$$

Using the shortfall expression, the echelon stock at installation i at the end of an arbitrary period is given by the random variable

$$\tilde{S}_i - Y_i - D(\lambda_i).$$

The distribution function of the shortfall and the echelon stock as well as the resulting average cost function can be analyzed by the concept of incomplete convolutions. Let G_1 and G_2 denote continuous distribution functions defined for nonnegative values. Then, G_2^Δ represents the distribution function of the random variable $(X_1 - \Delta)^+$.

---

[8] VAN HOUTUM, G.J., ZIJM, W.H.M. [1991].

$$G_2^\Delta(x) := \begin{cases} G_2(x + \Delta) & \text{if } x \geq 0 \\ 0 & \text{otherwise.} \end{cases}$$

The distribution $G_2^\Delta * G_1$, where $*$ is the convolution operator, denotes the incomplete convolution defined by

$$(G_2^\Delta * G_1)(x) = \int_0^x G_2(x + \Delta - u)dG_1(u), \qquad x \geq 0$$

which is the distribution function of the random variable $(X_1 - \Delta)^+ + X_2$.

Let F denote the single period demand distribution and F_λ be the corresponding λ-fold convolution that represents the demand distribution of λ consecutive time periods. Then, the distribution function of the shortfall is given by

$$\tilde{F}^{[S_1,...,S_i]}(x) = ((...((F_{\lambda_1}^{\Delta_1} * F_{\lambda_2})^{\Delta_2} * ... * F_{\lambda_{i-1}})^{\Delta_{i-1}})(x)$$

with $\Delta_i = \tilde{S}_i - \tilde{S}_{i+1}$ and $\tilde{S}_i = \min\{S_1, ..., S_i\}$. The distribution function of the echelon stock at i at the end of an arbitrary period (given by the desired order-up-to-level minus shortfall minus demand during the last λ_i periods) is given by

$$F^{[S_1,...,S_i]}(x) = ((...((F_{\lambda_1}^{\Delta_1} * F_{\lambda_2})^{\Delta_2} * ... * F_{\lambda_{i-1}})^{\Delta_{i-1}}) * F_i)(x).$$

For the computation of optimal base stock levels in a serial system, the average cost D_i introduced in the previous section can be written in terms of incomplete convolutions.[9] Recalling that D_i is defined as the sum of average cost of stockpoint i and its successors given that the supplier of stockpoint i can always deliver, the shortfall random variable and the corresponding distribution have to be adjusted to this condition. Then, for every stockpoint j situated downstream to the stockpoint i under consideration, the adjusted order-up-to-level, the shortfall, and the respective distributions are defined by:

$$\tilde{S}_j = \min\{S_i, ..., S_j\}$$
$$\Delta_j = \tilde{S}_j - \tilde{S}_{j+1}$$
$$Y_j = (...(D(\lambda_i) - \Delta_i)^+ + ... + D(\lambda_{j-1}) - \Delta_{j-1})^+$$
$$\hat{F}^{[S_i,...,S_j]} = (F_{\lambda_i}^{\Delta_i} * ... * F_{\lambda_{j-1}})^{\Delta_{j-1}}$$
$$F^{[S_i,...,S_j]} = (F_{\lambda_i}^{\Delta_i} * ... * F_{\lambda_{j-1}})^{\Delta_{j-1}} * F_{\lambda_j}.$$

Then, the conditional average cost function D_i becomes

$$D_i(S_i, ..., S_n) = \sum_{j=i}^n h_j^e \int_0^\infty (\tilde{S}_j - u)dF^{[S_i,...,S_j]}(u)$$

$$+ (p + h_n) \int_{\tilde{S}_n}^\infty (u - \tilde{S}_n)dF^{[S_i,...,S_n]}(u).$$

---

[9] VAN HOUTUM, G.J., ZIJM,W.H.M. [1997].

The partial derivative of the average cost function D_i with respect to S_i is given by

$$\frac{\partial}{\partial S_i} D_i(S_i, ..., S_n) = \sum_{j=i}^{n} h_j^e - (p + h_n)(1 - F^{[S_i, ..., S_n]}(\tilde{S}_n))$$

$$- \sum_{j=i+1}^{n} \hat{F}^{[S_j, ..., S_i]}(0) \frac{\partial}{\partial S_j} D_j(S_j, ..., S_n).$$

Applying the decomposition result, the term with the second sum vanishes for already determined optimal levels $S_j = S_j^*$, and the optimal order-up-to-level of stockpoint i is chosen to satisfy

$$F^{[S_i^*, ..., S_n^*]}(\tilde{S}_n) = \frac{p + \sum_{j=1}^{i-1} h_j^e}{p + h_n} \tag{4.7}$$

with $\tilde{S}_n = \min\{S_i^*, ..., S_n^*\}$. A bisection procedure can be applied to find the values $(S_i)_{i=n,...,1}$ that satisfy (4.7). The numerical problem arises from the computation of incomplete convolutions. In general, incomplete convolutions have to be evaluated by numerical integration. For large systems this becomes intractable. Therefore, BADINELLI[10] develops polynomial approximations. For Mixed-Erlang distributed demand, VAN HOUTUM, ZIJM[11] derive exact expressions for incomplete convolutions. Though these expressions can be evaluated in reasonable computation time, a small coefficient of variation causes a large number of Erlang phases. In this case, more computation time is required and very often numerical problems occur. SEIDEL, DE KOK [12] propose an iterative moment approximation method. Given the first two moments of X_1, a Mixed-Erlang distribution can be fitted as already presented for single-echelon models. Expected value and variance of $(X_1 - \Delta)^+$ are given by

$$E[(X_1 - \Delta)^+] = \frac{\kappa}{l}\left(1 - ME_{\kappa+1,q,l}(\Delta)\right) - \Delta\left(1 - ME_{\kappa,q,l}(\Delta)\right)$$

$$- \frac{q}{l}\left(1 - E_{\kappa,l}(\Delta)\right)$$

$$Var[(X_1 - \Delta)^+] = \frac{\kappa^2}{l^2}\left(1 - ME_{\kappa+2,q,l}(\Delta)\right) - \frac{2\Delta\kappa}{l}\left(1 - ME_{\kappa+1,q,l}(\Delta)\right)$$

$$+ \Delta^2\left(1 - ME_{\kappa,q,l}(\Delta)\right) + \frac{\kappa(1-q)}{l^2}\left(1 - E_{\kappa+2,l}(\Delta)\right)$$

$$- \frac{\kappa q}{l^2}\left(1 - E_{\kappa+1,l}(\Delta)\right) + \frac{2\Delta q}{l}\left(1 - E_{\kappa,l}(\Delta)\right)$$

$$- (E[(X - \Delta)^+])^2 \quad .$$

---

[10] BADINELLI, R.D. [1996].
[11] VAN HOUTUM, G.J., ZIJM, W.H.M. [1997].
[12] SEIDEL, H.P., DE KOK, A.G. [1990].

Then, mean and variance of $(X_1 - \Delta)^+ + X_2$ are given by $E[(X_1 - \Delta)^+] + E[X_2]$ and $Var[(X_1 - \Delta)^+] + Var[X_2]$ respectively (note that X_1 and X_2 are independent random variables) and again a Mixed-Erlang distribution can be fitted on these two values. This method is repeated until the value of the incomplete convolution is obtained.

4.1.1.3 Models with Service Level Constraints.

As outlined in VAN HOUTUM, INDERFURTH, ZIJM[13], the service level that is achieved by echelon order-up-to-levels $(S_i)_{i=1,...,n}$ at the final installation of a serial system can be expressed by incomplete convolutions.

The α-service level that expresses the non-stockout probability of the production/inventory system at the end of an arbitrary period is given by the probability that the sum of final-stage shortfall Y_n and demand during the last λ_n periods does not exceed the final-stage order-up-to-level S_n.

$$\alpha = P\{Y_n + D(\lambda_n) \leq S_n\}$$
$$= F^{[S_1,...,S_n]}(S_n).$$

The stockout size related service measure β necessitates the evaluation of the expected backlog of the final stage at the end and at the beginning of an arbitrary period. The expected backlog at the beginning of a period is given by the expected overshot of the final-stage shortfall and $\lambda_n - 1$ period demand over the respective order-up-to-level S_n. The expected backlog at the end of a period similarly results from the overshot of the shortfall and λ_n periods demand over S_n. Inserting these expressions into the definition of the β-service level yields

$$\beta = 1 - \frac{1}{\mu} \cdot \left(\int_{S_n}^{\infty} (u - S_n) d\hat{F}^{[S_1,...,S_n]} * F_{\lambda_n+1}(u) \right.$$
$$\left. - \int_{S_n}^{\infty} (u - S_n) d\hat{F}^{[S_1,...,S_n]} * F_{\lambda_n}(u) \right).$$

For the γ-service level, only the expected final-stage backlog at the end of a period is relevant.

$$\gamma = 1 - \frac{1}{\mu} \cdot \int_{S_n}^{\infty} (u - S_n) d\hat{F}^{[S_1,...,S_n]} * F_{\lambda_n+1}(u).$$

In applications where expected inventory holding costs have to be minimized subject to a service level constraint (instead of minimizing expected inventory holding and penalty costs), the above service level expressions apply together with the penalty cost solution approach. In situations with a predetermined stockout probability, the corresponding penalty cost parameter p that is required for the algorithm results from observing the equivalence relation[14]

---

[13] VAN HOUTUM, G.J., INDERFURTH, K., ZIJM, W.H.M. [1996].

[14] Note that this is the same equivalence mentioned for the single-echelon newsboy problem.

$$\alpha = F^{[S_1,\dots,S_n]} = \frac{p}{p + h_n}.$$

For β- and γ-service level constraints, no analytic equivalence relation can be utilized. Nevertheless, a penalty cost parameter that achieves the required service level will always exist but has to be determined numerically by iteratively applying the algorithm of VAN HOUTUM, ZIJM and finding the required p by bisection.

From a management accounting point of view, the evaluation of stockpoint and system performance with service levels is better suited for practical implementation than a penalty cost approach. For a given operating policy and demand realizations for the last periods, realized penalty costs are not as intuitive as realized service levels (whatever type of definition is appropriate for the underlying application context). Therefore, setting up a service level constraint for stockpoint and system performance enables or improves performance control. Nevertheless, from a theoretical point of view it is also possible to compare expected and realized shortage costs. In order to observe the origin of deviations in system performance, it is helpful to have internal service levels for each installation. For every given policy parameter set $(S_1, \dots, S_n)$, the implicit internal service levels can be obtained in the same manner as the system service levels.

$$\alpha_i = P\{Y_i + D(\lambda_i) \leq \Delta_i\} = F^{[S_1,\dots,S_i]}(\Delta_i) \; i = 1, \dots, n-1$$
$$\gamma_i = 1 - \frac{1}{\mu} \cdot \int_{\Delta_i}^{\infty} (u - \Delta_i) dF^{[S_1,\dots,S_i]}(u) \quad i = 1, \dots, n-1$$

In a periodic review system, ordering takes place at the beginning of the review period and there are no requirements during this review period except for final-stage demands. Therefore, the backlog at the end and at the beginning of a period are identical and the β-service level is meaningless for internal service level control.

4.1.2 No Delay Approaches

4.1.2.1 The Model of Simpson. In contrast to CLARK, SCARF, the approach of SIMPSON[15] follows a different assumption concerning the impact of internal shortages. System dynamics behind the SIMPSON model refer to the base-stock concept of KIMBALL[16]. Internal demands that are created by replenishment orders of downstream stockpoints are classified into two categories. These are separated by the maximum reasonable demand level which is set by management and depends on available capacity and flexibility to cope with extraordinary large demand realizations. Coverage for uncertain demand fluctuations below the maximum reasonable demand level is achieved by safety stocks. Emergency activities for non-regular internal

---

[15] SIMPSON, K.F. [1958].
[16] KIMBALL, G.E [1988]. See Section 3.2.2 for a description.

orders are necessary and assumed to be present in the SIMPSON approach. This second mode for dealing with large demands is called operating flexibility[17]. Extending the control of materials flow from a strategic perspective to an operative view, activities of production scheduling and order processing come into perspective. Rescheduling opportunities for planned production lots with higher priorities for orders that contain backlogged materials are available as well as accelerated (higher production intensity) or overtime production. These opportunities of operating flexibility are concerned with the backlogging assumption, only the material is delivered earlier than under the backorder case. An emergency supply of missing items from an external supplier corresponds to lost sales inventory control. In the following, it is assumed that operating flexibility refers to backorder related activities. Lost sales considerations are addressed in the extensions.

This two mode approach in inventory control was used by SIMPSON to develop a model for determining safety stock norms in a serial system. The approach is based on a direct linkage of stocks to time. Every safety stock level at a stockpoint suffices to deal with uncertain demand below the maximum reasonable level for some period of time called the coverage time. Depending on the available inventory, every material request is met after a service time ST_i.[18] This service time consists of the time required to have all components available (ST_{i-1}) plus the processing time λ_i it takes until products have been processed. The sum $ST_{i-1} + \lambda_i$ denotes the replenishment lead time of stockpoint i. Concerning the connection of the serial inventory system to the environment it is assumed that the external supplier of the first-stage stockpoint can always deliver ($ST_0 = 0$) and that the system produces to stock in order to meet all customer demands at the predetermined service level, that is there is no planned delay in customer request satisfaction ($S_n = 0$).

The goal of the SIMPSON approach is to find the optimal combination of service times (and equivalently the corresponding safety stock levels) that minimizes inventory holding costs on the one hand and guarantees a predetermined non-stockout probability (α-service level) on the other hand.

In the following, system dynamics of information and materials flow discussed for the model of KIMBALL in Section 3.2.2 are formalized. Every demand realization d_t that is observed at the beginning of the next review period $t + 1$ is immediately transmitted to all upstream stockpoints.

$$x_{i,t+1} = d_t \qquad i = 1, ..., n.$$

If every stockpoint i was able to deliver the requested products within its service time ST_i, planned net stock $\hat{y}_{i,t}$ at the end of a period and outstanding orders $\hat{q}_{i,t-j}$ for a given ordering policy with base-stock levels B_i and service

[17] GRAVES, S.C. [1988], INDERFURTH, K. [1991].
[18] For simplicity of the system dynamics description and without loss of generality, ST is assumed to be integer.

times ST_i are given by

$$\hat{y}_{n,t} = B_n - \sum_{j=0}^{\lambda_n + ST_{n-1}} d_{t-j}$$

$$\hat{y}_{i,t} = B_i - \sum_{j=1}^{\lambda_i + ST_{i-1}} d_{t-j} + \sum_{j=1}^{ST_i} d_{t-j} = B_i - \sum_{j=ST_i+1}^{\lambda_i + ST_{i-1}} d_{t-j} \qquad i = 1, ..., n-1.$$

At the final-stage stockpoint, demands during the lead time of $\lambda_n + ST_{n-1}$ have not arrived and the demand of the review period additionally lowers net stock. The other stockpoints $i = 1, ..., n-1$ face outstanding orders being identical to the demand of the previous $\lambda_i + ST_{i-1}$ periods. But they have not supplied the successor with the demands $(d_{t-1}, ..., d_{t-ST_i})$ during the service time ST_i. Given that no additional delays are introduced by a stockpoint, the service time cannot exceed the replenishment lead time. Therefore, feasible service times for stockpoints $i = 1, ..., n-1$ vary between 0 and $\lambda_i + ST_{i-1}$.

$$0 \leq ST_i \leq ST_{i-1} + \lambda_i \qquad i = 1, ..., n-1$$

The pipeline to stockpoint i at the end of period t is characterized by shipments

$$\hat{q}_{i,t-j} = d_{t-j-ST_{i-1}} \qquad j = 1, ..., \lambda_i.$$

The resulting system dynamics for net stocks are

$$\hat{y}_{n,t} = \hat{y}_{n,t-1} + \hat{q}_{n,t-\lambda_n} - d_t$$
$$\hat{y}_{i,t} = \hat{y}_{i,t-1} + \hat{q}_{i,t-\lambda_i} - \hat{q}_{i+1,t} \qquad i = 1, ..., n-1.$$

The condition that every order, even those exceeding the maximum reasonable demand level, is completely filled within the service time, requires for an operative organization of flexibility use. An internal delay is present if real net stock $y_{i,t}$ or an outstanding order $q_{i,t-j}$ deviates from the projected value $\hat{y}_{i,t}$ or $\hat{q}_{i,t-j}$ respectively. In order to ensure the no delay assumption, processing or delivery of products has to be speeded up, if there is not sufficient on-hand stock at stockpoint $i-1$ to start the complete processing of $\hat{q}_{i,t}$. The shipment quantity to stockpoint i at the beginning of period t is given by

$$q_{i,t} = \min\{OH_{i-1,t-1} + q_{i-1,t-\lambda_{i-1}+1}; \hat{q}_{i,t}\}$$

with $OH_{0,t} = \infty$ (assuming that the external supplier can always deliver). The resulting shortage of stockpoint $i-1$ to meet the demand within the service time ST_{i-1} is given by

$$b_{i-1,t} = (q_{i-1,t} - \hat{q}_{i-1,t})^+ \qquad i = 2, ..., n.$$

Therefore, the quantity $b_{i-1,t}$ has to be speeded up from pipeline stock of stockpoint $i-1$ or even from physical and pipeline stock of predecessors of $i-1$ at the beginning of period t.[19] For the speeding up procedure to be described in this section, two additional assumptions are made.

- FIFO priority in speeding up
  The quantities $b_{i-1,t}$ are picked from outstanding replenishment orders in the sequence of placement. This means that an older order in the pipeline is speeded up first.
- Only the missing quantity $b_{i-1,t}$ is picked.
  This implies that outstanding orders can be split.

The following algorithm describes the speeding up to meet the no delay assumption.

for $i := n - 1$ **downto** 1 **do begin**
  $b_{i,t} := (q_{i,t} - \hat{q}_{i,t})^+$
  $l := i$
  while $b_{i,t} > 0$ **do begin**
    $j := \lambda_l$
    repeat
      $q_{l,t-j+1} := (q_{l,t-j+1} - b_{i,t})^+; \; b_{i,t} := (b_{i,t} - q_{l,t-j+1})^+; \; j := j - 1$
    until $j = 0$ **or** $b_{i,t} = 0$
    if $b_{i,t} > 0$
      then $l := l - 1; \; b_{i,t} := (b_{i,t} - OH_{l,t-1})^+;$
        $OH_{l,t-1} := (OH_{l,t-1} - b_{i,t})^+$
  end;
end;

Note that the algorithm always terminates with $b_{i-1,t} = 0$ because even the external order released just before speeding up can be made available. As an extension to external requirements, shortages can be avoided by speeding up the final stockpoint's pipeline. Then, the algorithm starts with $i = n$ instead of $n - 1$.

The resulting information and materials flow is illustrated by using the example introduced for the CLARK, SCARF model. Base-stock levels are set to $B_1 = 25$ and $B_2 = 35$ which initiates the same information flow as in the CLARK, SCARF model. Service times are set to $ST_1 = ST_2 = 0$. This indicates that safety stocks with sizes of $SS_i = 5$ are held at both installations. The resulting stocks and shipment quantities are presented in Table 4.2. Deviations that result from the use of operating flexibility in cases where service times cannot be realized by the available amount of physical stock are indicated by adding/subtracting the number of items that are sped up. As

---

[19] In order to specify a sequence of events, speeding up takes place just after order release.

Table 4.2. Materials coordination example for the SIMPSON model

t	d_t	Stage 1				Stage 2			
		$q_{1,t}$	$q_{1,t-1}$	$OH_{1,t}$	$x_{1,t+1}$	$q_{2,t}$	$q_{2,t-1}$	$y2,t$	$x_{2,t+1}$
0	-	10	10	5	10	10	10	5	10
1	15	10	10	5	15	10	10	0	15
2	15	15	10	0	15	15	10	-5	15
3	5	15	15-5	0	5	10+5	15	0	5
4	5	5	15	5	5	5	15	10	5
5	20	5	5	15	20	5	5	5	20
6	10	20	5	0	10	20	5	0	10
7	5	10	20-5	0	5	5+5	20	0	5
8	5	5	10	10	5	5	10	15	5

already outlined in the previous section, internal stock insufficiencies occur in period 3 and 7. In order to guarantee the service time of zero periods from stockpoint 1 to 2, i.e. to deliver the requested products immediately, parts of pipeline orders are speeded up. The stock insufficiency in period 3 is covered by immediately making 5 items available from the order of 15 that would regularly arrive at the beginning of period 4. In period 7, 5 items are pushed from the order of 20 that would regularly arrive at the beginning of period 8.

In some cases, this kind of flexibility may not be available. As a consequence, order processing cannot start in time. Nevertheless, if processing starts $j = 1, ..., \lambda_i$ periods later, the result of the no delay assumption is not affected as long as the processing time can be reduced by j periods so that the order arrives at stockpoint i at the same time as it would have arrived if the material would have been available λ_i periods ago. Within this weakened no delay assumption, present net stock $y_{i,t}$ at the end of a period is compared to planned net stock $\hat{y}_{i,t}$. The speeding up quantity is now assumed to be picked at the end of the period t:

$$b_{i,t} = (y_{i,t} - \hat{y}_{i,t})^+ \qquad i = 1, ..., n.$$

The speeding up algorithm is easily adapted by inserting this new quantity. The materials flow is illustrated in Table 4.3. In comparison to the materials flow of the original interpretation of the no delay assumption, emergency actions are now taken at stockpoint 2 where processing times are reduced instead of speeding up the pipeline of stockpoint 1. Delivery lateness at stockpoint 1 creates a backlog which would not be present if a service time of zero periods could be guaranteed. Instead of speeding up at the beginning of period 3 and 7 (to start production at stage 2 with the desired quantity), now production of process 2 ends with the desired arrival quantities at the end of period 4 and 8 respectively. In order to shorten future speeding up requirements it can be tried to use present flexibility before emergency actions have to be carried out. Therefore, it can be checked if speeding up is

Table 4.3. Materials coordination example for the SIMPSON model (weak version)

		Stage 1				Stage 2			
t	d_t	$q_{1,t}$	$q_{1,t-1}$	$OH_{1,t}$	$BL_{1,t}$	$x_{1,t+1}$	$q_{2,t}$	$q_{2,t-1}$	$y_{2,t}$	$x_{2,t+1}$
0	-	10	10	5	0	10	10	10	5	10
1	15	10	10	5	0	15	10	10	0	15
2	15	15	10	0	0	15	15	10	-5	15
3	5	15	15	0	5	5	10	15	0	5
4	5	5	15	5	0	5	10-5	10+5	10	5
5	20	5	5	15	0	20	5	5	5	20
6	10	20	5	0	0	10	20	5	0	10
7	5	10	20	0	5	5	5	20	0	5
8	5	5	10	10	0	5	10-5	5+5	15	5

necessary in the future (because of delays in the past) and whether this can be performed in the present period or not.

After the illustration of the materials flow and the organization of information processing for a system with given order-up-to-levels (and implicitly given safety stocks), the determination of optimal safety stocks at each stockpoint is outlined in the following. With respect to the data required for the SIMPSON model, the following assumptions are made.

(1) The network is of serial type. Numbering of stockpoints is according to the notation introduced in 3.1.
(2) External customer demands are assumed to be normally distributed with expected value μ and standard deviation σ. Unsatisfied customer demands are backordered.
(3) Processing times for each stockpoint are described by λ_i. The final-stage processing time λ_n includes the review period.
(4) The cost structure only includes linear inventory holding costs h_i. Backorders are limited by a non-stockout probability constraint with service level α_n.
(5) Internal service levels α_i, $i = 1, ..., n - 1$ express the level of maximum reasonable demand.
(6) The external supplier can always deliver any requested amount of material.
(7) Time data (processing time parameter and service time decision variables) is assumed to be continuous. This simplifies calculations and can be justified by considering time averages.

The basic idea behind the approach is that delivery times and stocks are related. The system of n serial stockpoints is decoupled into separate stock keeping units where every unit can hold some amount SS_i of safety stock. The connection of stockpoints is achieved by expressing the time delay from one stockpoint to its immediate successor. Recalling the safety stock formula for single-echelon inventory systems, every amount of safety stock SS_i at installation i corresponds to a replenishment lead time. This corresponding time is called safety stock coverage time T_i.

$$SS_i = k_i \cdot \sigma_i \cdot \sqrt{T_i} \qquad (4.8)$$

As outlined for single-echelon models, the safety factor k_i for an α-service level is independent of the coverage time and is given by the α_i-percentile of the standard normal distribution function, i.e. $k_i = \Phi^{-1}(\alpha_i)$. The amount SS_i suffices to cover maximum reasonable demand over a time span of T_i periods. Connecting this equivalence of coverage time and safety stock with the presence of sufficient operating flexibility, every internal order released by installation $i + 1$ can be delivered by installation i after a service time ST_i.

The replenishment lead time of a stockpoint, the time required to make an item of product i available for use, is given by the service time of predecessor $i - 1$, which represents the time to make the component available for production plus the processing time λ_i. Taking into account a safety stock coverage time of T_i periods, the service time guaranteed by stockpoint i to its successor $i + 1$ is given by the difference between replenishment lead time and coverage time. Therefore, the service time of a stockpoint can never exceed the sum of predecessor service time plus the processing time of i.

$$ST_i = ST_{i-1} + \lambda_i - T_i \qquad i = 1, ..., n. \qquad (4.9)$$

The safety stock SS_i absorbs reasonable demand variations in the last T_i periods of the replenishment lead time while the successor has to deal with demand fluctuations during the first ST_i periods. The service time ST_0 of the external supplier is equal to zero due to the fact that the external supplier is assumed to be able to deliver any demanded amount of stock within the supply lead time of λ_1 periods. Because of the make-to-stock environment with the consequence of immediate delivery to customers, $ST_n = 0$ has to hold. This implies that the final stage stockpoint always covers its lead time $ST_{n-1} + \lambda_n$.

With the objective to minimize safety stock holding costs, the following non-linear optimization problem has to be solved.[20]

$$\min \ C = \sum_{i=1}^{n} h_i \cdot \sigma_i \cdot k_i \cdot \sqrt{ST_{i-1} + \lambda_i - ST_i}$$

$$\begin{aligned}
s.t. \qquad ST_i &\leq ST_{i-1} + \lambda_i & i = 1, ..., n-1 \\
ST_0 &= 0 \\
ST_n &= 0 \\
ST_i &\geq 0 & i = 1, ..., n-1.
\end{aligned}$$

Holding costs induced by safety stocks that are required in order to guarantee service times ST_i for reasonable demand situations are minimized under the

[20] SIMPSON, K.F. [1958].

constraints that service times are non-negative, continuous (see assumption (7)) variables and that a stockpoint's service time cannot exceed its replenishment lead time. Instead of modeling the problem in terms of service times, the optimization problem can be equivalently expressed by coverage times. Substitution of

$$T_i = ST_{i-1} + \lambda_i - ST_i$$

leads to the following optimization problem formulation.

$$\min \ C = \sum_{i=1}^{n} h_i \cdot \sigma_i \cdot k_i \cdot \sqrt{T_i}$$

$$s.t. \quad \sum_{j=1}^{i} T_j \leq \sum_{j=1}^{i} \lambda_j \qquad i = 1, ..., n-1$$

$$\sum_{j=1}^{n} T_j = \sum_{j=1}^{n} \lambda_j$$

$$T_i \quad \geq \quad 0 \qquad\qquad i = 1, ..., n.$$

Holding costs that result from safety stocks implemented by coverage times T_i are minimized under the constraints that cumulative coverage times cannot exceed cumulative processing times, whereas the total cumulative processing time has to be covered.

Both optimization problems are of non-linear type. A concave objective function[21] is minimized under linear constraints. For this class of optimization problem, an extreme point property holds. This is also true if on-hand stock inventory holding costs are minimized instead of safety stock holding costs.[22] In this case, the objective function of the coverage time optimization problem is given by

$$\min C = \sum_{i=1}^{n} h_i \cdot \sigma \cdot (k_i \cdot \Phi(k_i) + \phi(k_i)) \cdot \sqrt{T_i}$$

with ϕ and Φ representing the standard normal density and the cumulative density respectively.

From the extreme point property, it follows that values for optimal service times or coverage times for the respective optimization problem result from the extreme points of the convex solution set that is spanned by the linear constraints. For the service time optimization problem, the decision values that result from the extreme points are[23]

$$ST_i^* \in \{0; ST_{i-1} + \lambda_i\} \qquad i = 1, ..., n-1.$$

---

[21] See Appendix A.1.1.
[22] See Appendix A.1.2.
[23] SIMPSON, K.F. [1958].

Because of the successive structure of constraints in the coverage time optimization problem, each $\leq$ constraint with index i is only affected by decision variables from stockpoints with index smaller or equal to i. These values can be characterized by a forward logic, that is every value for a stockpoint i only depends on the chosen values for upstream stockpoints.

$$T_n^* = \lambda_n + \sum_{i=1}^{n-1}(\lambda_i - T_i^*),$$

$$T_i^* \in \left\{0; \lambda_i + \sum_{j=1}^{i-1}(\lambda_j - T_j^*)\right\} \qquad i = 1, ..., n-1.$$

The final stockpoint covers its replenishment lead time which is identical to the cumulative uncovered processing times of the entire serial system. For the remaining $n - 1$ stockpoints, the implication of the extreme point property is two-fold. First, a stockpoint coverage time is either zero or equal to the replenishment lead time. The corresponding safety stock is either equal to zero or it is sufficient to cover regular demand variability within the replenishment lead time. The second implication directly follows from the first. If the coverage time is positive, it always covers demand variations during a sum of processing times of consecutive stockpoints and simultaneously, coverage times for these stockpoints are zero. SIMPSON calls this result a stockpoint consolidation effect. With respect to materials coordination of the optimal safety stock policy, only some stockpoints will hold positive inventory, whereas other immediately forward the incoming material. Therefore, from a theoretical point of view, these stockless inventory points together with the next stock keeping installation can be consolidated to a single stockpoint with a processing time that is equal to the sum of the corresponding individual processing times. This consolidated inventory system will induce the same materials flow. Even regarding the SIMPSON model as an approximate tool, it will yield at least good candidates for stockpoint consolidation in a design phase for the inventory system.[24]

The stockpoint consolidation aspect becomes more evident if an equivalent backward characterization for optimal coverage times is derived from the forward scheme. Starting with the most downstream stockpoint, an optimal coverage time for installation i can be represented by solely using the optimal values of downstream stockpoints.

$$T_i^* \in \begin{cases} \{0\} & \text{if } \sum_{j=i+1}^{n}(T_j^* - \lambda_j) > 0 \\ \left\{\sum_{j=l}^{i}\lambda_i, \, l = 1, ..., i\right\} & \text{if } \sum_{j=i+1}^{n}(T_j^* - \lambda_j) = 0 \end{cases} \qquad i = 1, ..., n-1.$$

---

[24] GRAVES, S.C. [1988].

Under this type of characterization, the coverage time depends on the cumulative excess coverage time of all downstream stockpoints. Only if excess downstream coverage is zero, a positive safety stock is necessary. In this case, stockpoints $j = l, ..., i$ are consolidated where l represents the most upstream stockpoint included in the consolidation. If excess downstream coverage is positive, the processing time of i has already been covered and the resulting coverage time is zero.

In a non-recursive formulation for coverage times, the set of all relevant values (disregarding connections) is given by

$$T_n^* \in \left\{ \sum_{j=l}^{n} \lambda_j, \, l = 1, ..., n \right\}$$

$$T_i^* \in \{0\} \cup \left\{ \sum_{j=l}^{i} \lambda_j, l = 1, ..., i \right\} \qquad i = 1, ..., n - 1.$$

Summarizing, the advantage of the SIMPSON approach is that it allows for a problem decomposition into a multi-stage allocation problem of coverage against regular uncertain customer demands. From the optimization it follows that n independent single-echelon safety stock planning models (with the optimal coverage times representing the lead times) are applied.

4.1.2.2 Dynamic Programming Algorithms. In general, both optimization problems presented in the previous section can be solved by techniques developed for concave minimization problems.[25] The successive structure of the constraints in the coverage time minimization problem together with the possibility to characterize candidates for optimal coverage times explicitly allows for the application of enumerative algorithms, here Dynamic Programming. The fundamental idea of Dynamic Programming[26] is to solve a simultaneous optimization problem sequentially by introducing successive decision stages i. The evaluation of the objective function requires separability within the decision variables $\underline{u}_i$ of different solution stages. At every stage of the solution process, one or more decisions have to be made. Because of the simultaneous impact of all decision variables, this decision is affected by decisions made on stages evaluated earlier and itself affects decisions on succeeding solution stages. The situation determined by previous decisions is given by the state $\underline{z}_i$ of stage i. For every possible state (given by the state space Z_i) that is represented by the values of the corresponding state vector, the optimal decision with respect to the objective criterion must be derived. All decisions $\underline{u}_i$ which are feasible under the state $\underline{z}_i$ are included in the decision space $U_i(\underline{z}_i)$. For the overall optimal policy, this decision is only preliminary because at this point of the decision process it is not known

---

[25] HORST, R., TUY, H. [1996], Part B.
[26] BELLMAN, R. [1957].

whether the stage will face this state within the optimal strategy. An optimal decision, given the state of the stage, is taken with respect to direct costs resulting from the decision plus costs of succeeding stages that are influenced by this decision.

This logic can be implemented two-fold, starting either with the first or the final stage. Starting with the final stage, the recursion process is called backward recursion. For the final stage, there are zero costs for succeeding stages and for the first stage, there is only a single state. Preliminary cost values for the succeeding stage are available and optimal decisions have to be generated for every possible state of stage i. The first stage decision determines the optimal policy which is derived by a forward calculation. When the solution process is implemented the other way around, starting with the first stage and ending with the final stage, this is called forward recursion. Within this Dynamic Programming variant, the state represents the effect of all decisions of succeeding stages and preliminary decisions and corresponding costs represent the effect on preceding decisions. It is obvious that both approaches are equivalent.

Utilizing this idea for solving the safety stock coverage time allocation problem reduces the solution effort considerably. Not all 2^{n-1} different policy combinations that directly follow from the extreme point property with 2 candidate values (cover all or nothing) for each stockpoint (except the final one) have to be evaluated explicitly. Some solution alternatives share a common coverage strategy up to a distinct stockpoint and because of separability of the objective function, this common part of a solution is only evaluated once.

Within both alternative variants of Dynamic Programming, the stages of iteration are identical to the stockpoints and decision variables u_i represent the coverage times T_i. For a backward recursion formulation, states z_i denote the replenishment lead time, that is the cumulative uncovered processing times of upstream stockpoints (including the processing time of i)

$$z_i = \lambda_i + \sum_{j=1}^{i-1}(\lambda_j - u_j) \qquad\qquad i = 1, ..., n.$$

The state transition equation denotes how the state of the succeeding stockpoint i depends on the state of i and the decision made by i. Therefore, the replenishment lead time of $i + 1$ is given by the replenishment lead time of i minus coverage time plus processing time of $i + 1$.

$$z_{i+1} = z_i + \lambda_{i+1} - u_i \qquad\qquad i = 1, ..., n - 1.$$

The starting value is given by $z_1 = \lambda_1$ because the delivery time of the external supplier cannot be influenced.

From the extreme point properties presented in the previous section, it is possible to characterize the state spaces Z_i and decision spaces $U_i(z_i)$ of all stockpoints explicitly.

$$Z_i = \left\{ \sum_{j=l}^{i} \lambda_j \ , \ l = 1, ..., i \right\} \qquad i = 1, ..., n$$

$$U_n(z_n) = \{z_n\}, \ U_i(z_i) = \{0 \ ; \ z_i\} \qquad i = 1, ..., n-1.$$

Depending on the last predecessor $l - 1$ of i that holds positive safety stock, the replenishment lead time is given by the sum of processing times from stockpoint l to i. The state dependent decision is identical to the state for the final-stage. For the other $n - 1$ stockpoints, it can only obtain values 0 or z_i.

Direct costs assigned at a stage can either be based on the safety stock or the on-hand stock holding cost criterion.

- Safety stock holding cost criterion

$$c_i(u_i) = h_i \cdot \sigma_i \cdot k_i \cdot \sqrt{u_i} \qquad i = 1, ..., n$$

- On-hand stock holding cost criterion

$$c_i(u_i) = h_i \cdot \sigma_i \cdot (k_i \cdot \Phi(k_i) + \phi(k_i)) \cdot \sqrt{u_i} \qquad i = 1, ..., n$$

Final-stage cost values for covering z_n time periods only depend on costs $c_n(z_n)$ assigned by the final stage.

$$f_n(z_n) = c_n(z_n) \qquad \forall z_n \in Z_n.$$

The value function $f_i(z_i)$ for stockpoints $i = 1, ..., n-1$ represents the optimal costs of stockpoint i and downstream stockpoints if the lead time of the stockpoint is equal to z_i and is derived from the functional equation

$$f_i(z_i) = \min_{u_i \in U_i(z_i)} \{c_i(u_i) + f_{i+1}(z_i + \lambda_{i+1} - u_i)\} \qquad \forall z_i \in Z_i.$$

The decision that minimizes the cost expression is the preliminary decision $u_i^*(z_i)$ that will be implemented if state z_i is reached under optimal decisions of all preceding stages. Beginning with $u_1^*(\lambda_1)$, the states of all stockpoints under optimal coverage can be computed by a forward calculation that inserts the preliminary optimal decisions into the state transition equation

$$z_{i+1}^* = z_i^* + \lambda_{i+1} - u_i^*(z_i^*) \qquad i = 1, ..., n-1.$$

The unconditioned optimal coverage decisions to be implemented are $u_i^* = u_i^*(z_i^*)$ and the corresponding safety stocks are given by inserting u_i^* into (4.8).

Instead of evaluating 2^{n-1} solutions that result from the extreme point properties, the presented formulation has a complexity of $O(n^2)$. Computation time that is required to perform the Dynamic Programming algorithm is influenced by the storage of preliminary decisions and cost values for each

state and by the number of cost expressions that have to be evaluated to find
the minimum of the value function. The state space of stockpoint i consists
of i possible lead times, whereas, the decision space for every state has a size
of 2 to find the cost minimum (except for the final installation with size 1 for
every state). Therefore, the total number of states is given by

$$\sum_{i=1}^{n} i = \frac{n}{2}(n+1)$$

whereas, the total number of evaluated decisions is

$$\sum_{i=1}^{n-1} 2i + n = n^2.$$

Concerning the objective criterion it should be noticed that the evaluation
of an on-hand stock criterion is more cumbersome due to the computation of
the standard normal density and cumulative density functions.

Equivalently, a forward recursion can be applied. Calculation starts with
evaluating the most upstream stockpoint. Using the optimal preliminary cost
of upstream stockpoints, all installations are evaluated until the most downstream stockpoint is reached. Thereby, the impact of downstream decisions
is included in the state variable of a stockpoint. The impact on upstream
decisions results from the value function. The definition of decision variables
u_i for each stockpoint is identical to the backward recursion. In contrast, the
state of a stockpoint reflects the cumulative excess coverage of stockpoints
being situated downstream to i. This is the cumulative time that is covered
by safety stocks held at stockpoints $i+1, ..., n$ in excess to corresponding
processing times, i.e. the time span of uncertain demands that has already
been covered by safety stocks at downstream installations.

$$z_i = \sum_{j=i+1}^{n} (u_j - \lambda_j) \qquad i = 1, ..., n-1.$$

The state transition equation describes the development of cumulative excess
coverage from stockpoint $i+1$ to i. Taking into account cumulative excess
coverage to $i+1$, the processing time λ_{i+1} must be covered in any event (by
safety stocks or from using excess downstream coverage).

$$z_i = z_{i+1} - \lambda_{i+1} + u_{i+1} \qquad i = 1, ..., n-1$$

State transition starts with $z_n = 0$ because no excess coverage is faced by the
final installation when immediate response to customer demands is assumed.
Using the backward characterization of the extreme point property, the state
spaces are

$$Z_n = \{0\}, \ Z_i = \{0\} \cup \left\{ \sum_{j=l}^{i} \lambda_j \ , \ l = 1, ..., i \right\} \qquad i = 1, ..., n - 1.$$

For an optimal policy, excess coverage that is faced by a stockpoint can either be zero or an accumulation of processing times on the supply chain to any predecessor l (including $l = i$). From this stockpoint consolidation aspect, it follows that the decision space $U_i(z_i)$ contains the single value zero if the processing time of i has already been covered. Because of the backward characterization of the extreme point property, this is equivalent to $z_i > 0$. If cumulative excess coverage is zero, the stockpoint is free to cover the cumulative processing times on the chain to any predecessor l.

$$U_i(z_i) = \begin{cases} \{\sum_{j=l}^{i} \lambda_j \ , l = 1, ..., i\} & \text{if } z_i = 0 \\ 0 & \text{if } z_i > 0 \end{cases} \qquad i = 1, ..., n.$$

For the first-stage stockpoint, a positive coverage time is only necessary if the processing time is larger than excess coverage of downstream stockpoints. Therefore, the first stage decision is

$$u_1 = (\lambda_1 - z_1)^+$$

with corresponding costs

$$f_1(z_1) = c_1(u_1) \qquad \forall z_1 \in Z_1.$$

Direct cost functions c_i are defined in the same manner as outlined for the backward algorithm. State dependent optimal costs of stockpoints $i = 2, ..., n$ result from

$$f_i(z_i) = \min_{u_i \in U_i(z_i)} \{c_i(u_i) + f_{i-1}(z_i + u_i - \lambda_i)\} \qquad \forall z_i \in Z_i.$$

Given excess downstream coverage, the optimal decision minimizes the sum of costs implemented by safety stocks of stockpoint i and optimal costs for the upstream part of the system.

The complexity of the forward algorithm is of the same order as in the backward algorithm. The total number of states is

$$\sum_{i=1}^{n-1}(1+i) + 1 = \frac{n}{2}(n+1).$$

In (one per stockpoint) states with zero excess coverage, more decisions (identical to the number of predecessors plus one) have to be evaluated. In positive excess states (for all stockpoints except for $i = n$) only a zero coverage decision has to be evaluated. Therefore, the number of decisions is

$$\sum_{i=1}^{n} i + \sum_{i=1}^{n-1} \sum_{j=1}^{i} 1 = n^2.$$

4.1.2.3 Local Search Heuristics. Though computational complexity of the optimal solution algorithm is only $O(n^2)$ and, therefore, there is no real need for heuristic algorithms, a principle for developing local search procedures is presented for the serial system and extended in a later section to networks of general structure where computational complexity increases considerably. Heuristic algorithms can be roughly classified into solution construction and solution improvement procedures. Both types use an encoding scheme for a solution. While construction algorithms successively assign values to variables of the problem characterization, with or even without regarding a cost criterion, improvement methods iteratively modify a given solution until no further improvement is possible or more generally, until a stopping criterion terminates the iteration process.

The characterization of optimal coverage time allocation schemes shares common features with optimal solution properties found for other problems in logistics. Two well known examples are the dynamic lot-sizing problem and the warehouse location problem, where local search implementations have successfully been applied.[27] For a fixed configuration of setup periods in the lot-sizing problem, production quantities are given by cumulative demands from this period to the next setup period t (excluding demand in t). Therefore, every lot-sizing policy can be encoded by a binary string. The same holds for the uncapacitated warehouse location problem. For any given configuration of locations, every customer is serviced by the location with the lowest transportation costs to satisfy the corresponding demand. Therefore, a solution is completely characterized by a binary string.

These ideas can directly be applied to the safety stock coverage time allocation problem. Every extreme point solution can be encoded by a binary string $\underline{b} = (b(1), ..., b(n-1))$ of length $n-1$. A "1" at position i of the string indicates that stockpoint i holds a positive amount of safety stock whereas "0" represents a stockless inventory point. Because of the "cover all or nothing" property and the connected stockpoint consolidation, the coverage times that follow from a given string can be computed as follows. Let $p(i)$ denote the next upstream installation to point i that holds safety stock.

$$p(i) = \max_{j=1,...,i-1} \{j|b(j) = 1\}$$

Then, the lead time of stockpoint i is determined by the sum of processing times from $p(i) + 1$ to i and the coverage time is

$$T_i = \begin{cases} \sum_{j=p(i)+1}^{i} \lambda_j & \text{if } b(i) = 1 \\ 0 & \text{if } b(i) = 0 \end{cases} \qquad i = 1, ..., n.$$

Because the final stage stockpoint always has to cover its lead time, $b_n = 1$.

---

[27] SALOMON, M., KUIK, R., VAN WASSENHOVE, L.N. [1993], SCHILDT, B. [1994].

Based on the binary representation, procedures for generating a starting configuration are necessary. Besides a random solution generation, more problem related construction principles come into mind. A first approach is to start with allocation patterns derived from simple rules of thumb proposed for safety stocks in MRP systems. Alternatively, the similarity to the dynamic lot-sizing problem provides approaches with cost consideration based successive assignment of values to variables.

- Random solution generation

$$b(i) = \begin{cases} 1 & \text{if } random(0,1) \le 0.5 \\ 0 & \text{otherwise} \end{cases} \qquad i = 1, ..., n-1$$

This method is quite general but neither problem nor cost oriented. Therefore, it only generates a starting solution for following improvement algorithms.

- End-item buffering

$$b_i = 0, \qquad\qquad i = 1, ..., n-1$$

This method is problem oriented in the sense that it provides the safety stock allocation pattern proposed in the majority of MRP applications, i.e. to hold safety stocks at the end item level exclusively.

- First- and final-stage buffering

$$b_1 = 1, \ b_i = 0 \quad i = 2, ..., n-1$$

- All-stage buffering

$$b_i = 1 \qquad i = 1, ..., n-1$$

- Successive assignment I
  This approach is both problem and cost related. Values to the safety stock coverage string are sequentially assigned and fixed, beginning with the most downstream and approaching to the most upstream installation by regarding the direct cost impact of the assignment. For the stockpoint under consideration it is checked, if coverage of λ_i at the next safety stock holding installation is cheaper than covering this amount at i.

```
begin
    j := n; i := n - 1; lt_j := λ_j;
    while i > 0 do begin
        if c_i(λ_i) + c_j(lt_j) < c_j(λ_i + lt_j)
            then b(i) := 1; j := i; lt_j := λ_i
            else b(i) := 0; lt_j := lt_j + λ_i
        i := i - 1;
    end;
end;
```

- Successive assignment II

  Allocation of coverage is highly influenced by cost degression as a result of the square root formula. While successive assignment I successively approaches from one stockpoint to its predecessor by considering the single processing time, this variant includes a rough look ahead component. It is checked if cumulative processing times from the first stage to stockpoint i can be covered cheaper at i or at the next downstream stockpoint with already fixed positive safety stock decision.

  begin

  $$j := n; i := n - 1; cum := \sum_{m=1}^{i} \lambda_m; lt_j := \lambda_j;$$

  while $i > 0$ **do begin**
    if $c_i(cum) + c_j(lt_j) < c_j(cum + lt_j)$
     then $b(i) := 1; j := i; lt_j := \lambda_i$
     else $b(i) := 0; lt_j := lt_j + \lambda_i$
    $i := i - 1; cum := cum - \lambda_{i+1}$
  end;

  end;

After an initial solution has been generated by a construction heuristic, simple changes of this given solution can yield better solutions. This approach is called local search. Based on the binary encoding scheme of a solution, a simple change is to switch a single bit of the string. The problem oriented interpretation of such a modification depends on the direction of modification.

- Introduce a safety stock holding stockpoint i

  In this case, the bit for stockpoint i is set from 0 to 1. For the old solution, stockpoint i is a stockless inventory point and demand uncertainty during the lead time of i is covered by the next downstream installation that holds safety stock. The new solution that is generated by the adjustment covers the lead time of i by creating safety stocks at i. Therefore, the coverage time of the next downstream safety stock holding installation is decreased by this amount of time.

  $$p(i) = \max_{j=1,\ldots,i-1} \{j|b(j) = 1\}$$
  $$s(i) = \min_{j=i+1,\ldots,n} \{j|b(j) = 1\}$$

Therefore, the following modifications are made to obtain a neighbor solution.

$$b(i) := 1 - b(i)$$

$$T_i := \sum_{j=p(i)+1}^{i} \lambda_j$$

$$T_{s(i)} := \sum_{j=i+1}^{s(i)} \lambda_j$$

$$\Delta C = c_i(T_i) + c_{s(i)}(T_{s(i)}) - c_{s(i)}(T_i + T_{s(i)})$$

- Delete a safety stock holding stockpoint i

  In this case, the bit for stockpoint i is set from 1 to 0. Therefore, the time covered by stockpoint i before modification has to be covered by the next downstream installation that holds safety stock.

$$b(i) := 1 - b(i)$$

$$T_{s(i)} := \sum_{j=p(i)+1}^{s(i)} \lambda_j$$

$$\Delta C := c_{s(i)}(T_{s(i)}) - c_i(T_i) - c_{s(i)}(T_{s(i)} - T_i)$$

$$T_i := 0$$

The quality of modifications depends on the sign of the cost difference ΔC. If $\Delta C < 0$, the modified solution implies an improvement. For the general search process of finding better solutions from a given solution, different local search strategies are available. A first choice for the search strategy concerns the number of evaluated modifications (neighbors) to a given solution. One extreme is to accept the first improving modification as a new actual solution (descent variant), whereas, the other extreme is to evaluate all possible modifications to the actual solution and to select the best one (steepest descent variant).

- Accept first improvement

  Within this approach, the sequence of modifications is important. The bit to be modified can be chosen randomly or by a fixed order.
- (a) Random selection order

$$i := random(1, n - 1)$$

- (b) Fixed selection order

  For selecting a stockpoint in a serial system, it is reasonable to check stockpoints in the order given by the supply chain. Therefore, only the direction, that is whether to start with the most upstream or with the predecessor of the most downstream stockpoint, has to be chosen.
- Accept best improvement

$$i := \arg\min_{j=1,\dots,n-1} \{\Delta C(j) | \Delta C(j) < 0\}$$

The local search procedure terminates when no further improvement is found by all modifications to the actual solution. Only for random sequence selection, where it is not assured that all neighbors are investigated, it might be favorable to terminate after investigation of a fixed number of neighbors without solution improvement.

4.1.2.4 Effects in Allocation of Safety Stocks. In this section, different effects of safety stock allocation in multi echelon systems are analyzed in more detail. For ease of presentation, a two-echelon serial model under the SIMPSON assumptions and a safety stock holding cost criterion is used. It is assumed that the safety factor is identical for all stockpoints $k_i = k$ and $a_{1,2} = 1$. The coverage time optimization model reduces to

$$\min C = h_1 \cdot k \cdot \sigma \cdot \sqrt{T_1} + h_2 \cdot k \cdot \sigma \cdot \sqrt{T_2}$$
$$T_1 \leq \lambda_1$$
$$T_1 + T_2 = \lambda_1 + \lambda_2$$
$$T_1, T_2 \geq 0.$$

First, the fact that safety stocks can be held at both installations, and that different safety stock combinations provide an identical customer service level, yields a substitution relation between safety stocks. If safety stocks of the final product installation are reduced, safety stocks of the upstream installation have to be increased as depicted in Figure 4.2.

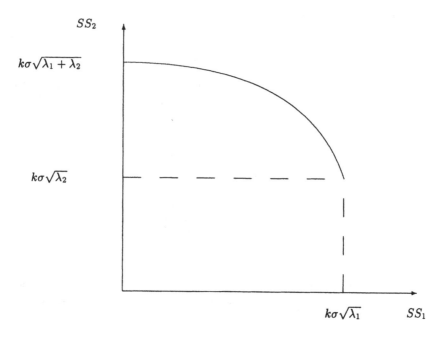

Fig. 4.2. Safety stock substitution in a two-echelon serial system

Using the objective function and solving for the safety stock of stockpoint 2, the depicted function is given by

$$SS_2 = k\sigma \sqrt{\lambda_1 + \lambda_2 - \left(\frac{SS_1}{k\sigma}\right)^2}.$$

Nevertheless, the pure substitution principle does not give any insight into the question whether safety stock allocation has any advantage in the sense of cost savings. A lot of early MRP literature advocates safety stocks only for the final echelon. Because of the square root safety stock relation, necessary safety stock increases with the square root if the replenishment lead time increases. By taking into account the pure square root effect, it is indeed optimal to hold safety stocks only at the final stage because

$$\sqrt{\lambda_1} + \sqrt{\lambda_2} \geq \sqrt{\lambda_1 + \lambda_2}.$$

The square root effect can be compensated by a value added effect. Value added to products by processing results in increasing inventory holding costs at downstream processing stages. Therefore, holding inventory is more expensive at downstream installations. The critical inventory holding cost ratio where end-item buffering and holding safety stocks at both stockpoints yield identical costs is given by

$$\frac{h_2}{h_1} = \frac{\sqrt{\lambda_1}}{\sqrt{\lambda_1 + \lambda_2} - \sqrt{\lambda_2}}.$$

If the ratio is smaller than the critical value, the square root effect dominates the value added effect and vice versa. The magnitude of the critical ratio is influenced by the lead times. The higher the lead time at the first-stage stockpoint, the lower the critical ratio. Note that under the assumption of

Table 4.4. Critical holding cost ratios

λ_1	1			5			10		
λ_2	2	6	11	2	6	11	2	6	11
h_2/h_1	3.15	5.10	6.78	1.82	2.58	3.27	1.54	2.04	2.50

identical safety factors k for both installations, these critical ratios are independent of the objective criterion, i.e. whether safety stock or on-hand stock is subject to holding costs.

4.1.3 Comparison and Synthesis

Both approaches presented in 4.1.1 and 4.1.2 are applied to a three-stage serial supply chain in order to demonstrate the numerical computations and to illustrate differences in the model outcome. Demand expectation and standard deviation are $\mu = 10$, $\sigma = 3$. Processing times for each stage are $\lambda_1 = 2$,

$\lambda_2 = 3$, and $\lambda_3 = 2$. Inventory holding costs with respect to a value added scheme are given by $h_1 = 1$, $h_2 = 3$, and $h_3 = 6$. Therefore, $h_1^e = 1$, $h_2^e = 2$, and $h_3^e = 3$ represent the corresponding echelon holding costs. The internal and external service levels of the SIMPSON model are set to $\alpha_i = 0.95$. The corresponding penalty cost required for the VAN HOUTUM, ZIJM algorithm in order to obtain a final-stage α-service level of 95% is $p = 114$.

Since the full delay approach of VAN HOUTUM, ZIJM is based on the assumption of Mixed-Erlang demand, the SIMPSON approach with its extreme point properties that are derived for normally distributed demands is applied to Mixed-Erlang demands, implicitly assuming that the extreme point property holds for this demand distribution as well. Table 4.5 contains the results of the backward Dynamic Programming algorithm with all relevant states, decisions in each state, and corresponding functional values. The optimal safety stock policy determined by forward calculus is $T_1^* = 2$, $T_2^* = 0$, and $T_3^* = 5$ for both demand distributions. Therefore, safety stocks are held at the first and at the final stage which implies that the second and the third stockpoint operate as a single installation with a lead time of five periods. These results rely on the assumption that every stockpoint can fulfill any

Table 4.5. Dynamic Programming results for the serial example

i	z_i	$u_i(z_i)$	Normal		Mixed-Erlang	
			$SS_i(u_i)$	$f_i(u_i)$	$SS_i(u_i)$	$f_i(u_i)$
3	2	2	6.98	41.88	7.45	44.70
	5	5	11.03	66.18	11.52	69.12
	7	7	13.06	78.36	13.55	81.30
2	3	0	0.00	66.18	0.00	69.12
		3	8.55	67.53	9.03	71.79
	5	0	0.00	78.36	0.00	81.30
		5	11.03	74.97	11.52	79.26
1	2	0	0.00	74.97	0.00	79.26
		2	6.98	73.16	7.45	76.57

order within the service time. If this no delay assumption does not hold, an operation with safety stocks determined by the SIMPSON approach will result in lower customer service levels than required, except for the solution alternative where the entire system is consolidated to a single stockpoint, that is that safety stocks are exclusively implemented at the final-stage stockpoint. The achieved non-stockout probabilities for internal and external stockpoints under Mixed-Erlang demands and safety stocks determined by the SIMPSON approach are shown in Table 4.6. In the present example, deviations are rather small. These values are obtained by calculating equivalent echelon order-up-to-levels by summing up base-stock levels and utilizing the service formulas presented in 4.1.1.3. The optimal echelon order-up-to-levels under the full delay assumption of the CLARK, SCARF model determined by the

Table 4.6. Full delay service levels of SIMPSON policy options

Solution			SS_i			B_i			$\hat{\alpha}_i$		
0	0	1	0	0	13.6	-	-	33.6	-	-	0.950
0	1	1	0	11.5	7.5	-	61.5	27.5	-	0.950	0.942
1	0	1	7.5	0	11.5	27.5	-	61.5	0.950	-	0.948
1	1	1	7.5	9.0	7.5	27.5	39.0	27.5	0.950	0.947	0.944

VAN HOUTUM, ZIJM algorithm result from the following three newsboy-type formulas.

$$F^{[S_3]} = \frac{117}{120} \Rightarrow S_3 = 29.13$$

$$F^{[S_2,S_3]} = \frac{115}{120} \Rightarrow S_2 = 63.90$$

$$F^{[S_1,S_2,S_3]} = \frac{114}{120} \Rightarrow S_1 = 86.80$$

Incomplete convolution expressions are evaluated by the moment fitting procedure of SEIDEL, DE KOK instead of utilizing cumbersome exact expressions. The optimal policy prescribes (local) safety stocks $SS_1 = 2.9$, $SS_2 = 4.77$, and $SS_3 = 9.13$ for all stockpoints. A simple valuation of these safety stocks with holding cost yields costs of $C = 71.99$. Compared to the Mixed-Erlang demand SIMPSON model result of $C = 76.57$, the full delay approach yields a better solution with respect to cost and service. The service superiority is obvious from the difference in the assumptions concerning internal delays. The cost inferiority mainly results from the high internal service levels of 95%. In order to allocate safety stocks in the SIMPSON model, each internal safety stock has to be sized with respect to a 95% non-stockout probability over the allocated coverage time. Therefore, internal buffers are rather expensive. On the other hand, the full delay model does not face this additional constraint when allocating buffers. The achieved internal service levels in the full delay solution are $\hat{\alpha}_1 = 0.7651$ and $\hat{\alpha}_2 = 0.873$. Assuming that these internal service levels would reflect the correct maximum reasonable demand in the SIMPSON model, the optimal safety stock policy of this approach would become $SS_1 = 2.9$, $SS_2 = 5.99$, $SS_3 = 7.45$ with $C = 65.57$.

Based on these insights obtained from a simple numerical example, the following subsections analyze (1) how do the SIMPSON safety stocks have to be increased in order to achieve the required service levels even under full delay conditions and (2) can the advantages of both approaches be combined by a joint method?

4.1.3.1 Adjustment of Safety Stock Levels in the Simpson Model.

If operating flexibility is not available in the manner assumed, an implementation with the determined safety stock levels will yield a lower level of service than desired. This service level deviation that results from base-stock levels

B_i when operating flexibility is not available can be computed by determining equivalent echelon-order-up-to-levels from the result of AXSÄTER, ROSLING and α-service level expressions of VAN HOUTUM, INDERFURTH, ZIJM. If the service undershot is significant, a tuning of policy parameters is possible by sequentially increasing safety stock levels. In order to keep the safety stock increment procedure as simple as possible, the following limiting assumptions are made.

- Safety stock increments are only analyzed for stockpoints with positive safety stocks. The stockpoint consolidation obtained from the coverage allocation optimization problem is preserved, therefore, no additional safety stock points are introduced.
- Safety stock adjustments are analyzed by two-stage service level interaction considerations. Increments follow the extreme strategies (a) to increase the upstream safety stock level or (b) to increase the downstream safety stock level, until the required downstream service level is reached.

Consider the set of safety stock holding stockpoints $\{i|SS_i > 0\}$ with corresponding coverage times T_i and resulting base-stock levels B_i. Proceeding with increasing stockpoint index, for every pair $(i, s(i))$ of a stockpoint and its next downstream stock holding successor, the realized downstream service level under no operating flexibility and given that no delay was introduced by predecessors of i is a function of the two base-stock levels B_i and $B_{s(i)}$.

$$\hat{\alpha}_{s(i)} = \int_0^{B_{s(i)}} F_{T_i}(B_i + B_{s(i)} - x) f_{T_{s(i)}}(x) dx = F^{[B_i + B_{s(i)}, B_{s(i)}]}(B_{s(i)})$$

Under Mixed-Erlang demands this expression can be analyzed as outlined in 4.1.1.2. For normally distributed demands, the analysis becomes more complex and numerical integration or approximations are required to evaluate the resulting service level.

In order to reduce the service level deviation $\alpha_{s(i)} - \hat{\alpha}_{s(i)}$, the upstream base-stock B_i (upstream increment) or the downstream base-stock $B_{s(i)}$ (downstream increment) is increased until the resulting service level deviation is zero or below a certain tolerance level. The respective amount of increment can be determined by a bisection procedure as outlined for the determination of order-up-to-levels. Both variants of the increment procedure yield additional safety stocks Λ_i for all stockholding installations except for the most upstream one (within the increase downstream variant) or except for the most downstream one (in the increase upstream variant). Therefore, the adjusted base-stock levels are

$$B_i^a = B_i + \Lambda_i.$$

The increment Λ_i is the additional safety stock required to cover internal delivery insufficiencies caused by stockpoint i (increase upstream variant) or by the stockholding predecessor $p(i)$ (increase downstream variant), given that insufficiencies caused by $p(i)$ or by $p(p(i))$ are negligible. In order to account for these more upstream effects, the calculation is always based on base stock levels B_i instead of B_i^a because the additional stock Λ_i serves for covering delays resulting from $p(i)$.

The outcome of both adjustment methods is illustrated using the previous example under Mixed-Erlang demands. Table 4.7 shows the additional safety stocks required to achieve service levels of 95% for all stockholding installations and for all possible coverage combinations within the downstream increment variant. The necessary adjustments are only minor and do not impact the optimal allocation choice although the rank order of the alternatives changes. The alternative increase upstream variant on the other hand is infe-

Table 4.7. Increase downstream adjustments

Solution			SS_i			Λ_2	Λ_3	$\hat{\alpha}_1$	$\hat{\alpha}_2$	C	C^{ia}
0	0	1	0	0	13.6	-	-	-	-	81.3	81.3
0	1	1	0	11.5	7.8	-	0.37	-	0.95	79.3	81.5
1	0	1	7.5	0	11.7	-	0.28	0.95	-	76.6	78.3
1	1	1	7.5	9.2	7.7	0.2	0.27	0.95	0.95	79.2	81.5

rior to the increase downstream method. Too large safety stock increments are required compared to the first variant that cannot be compensated by holding cost advantages. The post optimal adjustment of safety stock levels in order

Table 4.8. Increase upstream adjustments

Solution			SS_i			Λ_i			$\hat{\alpha}_i$		
0	1	1	0	23.0	7.5	-	11.5	-	-	0.95	0.95
1	0	1	14.0	0	11.5	6.6	-	-	0.95	-	0.95
1	1	1	14.0	18.0	7.5	6.6	9.0	27.5	0.95	0.95	0.95

to guarantee service levels under full delay conditions has the disadvantage that the overall optimal allocation after adjustment may be different to the allocation obtained under the no delay assumption. Especially the end-item buffering alternative is discriminated within this sequence of evaluation since end-item buffering never requires adjustments, whereas all other allocation solutions do. Therefore, it might be desirable to include the adjustment cost consequences into the Dynamic Programming or the local search procedures. The integration into the presented backward Dynamic Programming framework requires the incorporation of an additional state variable. In order to determine the additional adjustment requirements, the base-stock of the next upstream stockholding installation has to be known as an additional state. Then, besides the safety stock decision, the two-level adjustment consideration can be added in case that the complete replenishment lead time is covered as a preliminary decision. The integration into the local search procedures is straightforward. If an additional safety stock installation is introduced (by setting the respective indicator variable to 1), the old adjustment of the next downstream safety stock holding successor has to be readjusted with respect to the new base-stock levels. If the stockpoint faces a safety stock holding predecessor, a selfadjustment is necessary as well. If a safety stock holding installation is consolidated (by setting the respective indicator variable to 0), the next downstream safety stock point has to recalculate its safety stock adjustment except for the case where the entire system is consolidated to a single stockpoint.

4.1.3.2 Combination of Full Delay and No Delay Approaches. The SIMPSON model restricts every stockpoint to deliver the requested material within the service time. Large orders are made available by operating flexibility. The capacity that enables this flexibility or the level of maximum reasonable demand determines internal service levels α_i. If available capacity is small and therefore a large amount of demand has to be covered by safety stocks, the internal service level will be high and safety stocks are expensive at this installation. If the internal service level is sufficiently large, safety stocks at this installation are too expensive and, as a consequence, demand uncertainty during the corresponding processing time is covered at one of the downstream stockpoints.

The CLARK, SCARF model assumes that every amount of demand variability has to be covered by safety stocks exclusively. In cases with sufficient slack capacity, unnecessarily large stocks are held. On the other hand, safety stocks can be allocated to stockpoints with low holding costs rather cheaply because the inability to operate under no delay conditions is solved by shortfalls that are covered by downstream stockpoints.

Combination of the advantages of both approaches within a local search framework that coordinates both models is straightforward to the ideas presented for local search methods so far. Each stockpoint can be encoded with respect to three instead of two states. The value $b(i) = 0$ indicates that the stockpoint holds no safety stocks and, therefore, is consolidated with the next installation. If $b(i) = 1$, the stockpoint holds safety stocks and operates with no delay to its successor, that is that each successor order is completely and immediately satisfied. A stockpoint that holds safety stocks but operates with delays in situations where successor requirements exceed available inventory is encoded by $b(i) = 2$. The final-stage stockpoint is generally encoded by $b(n) = 1$. The evaluation of a given solution configuration follows the combined lines of both approaches and methods. It depends on the operation logic of the last stock holding predecessor j. If $b(j) = 1$, every order arrives without delay and safety stock requirements for i (regardless of $b(i) = 1$ or $b(i) = 2$) are determined from the SIMPSON approach. Contrary, $b(j) = 2$ implies that i faces a shortfall induced by delivery insufficiencies of j and therefore, safety stock requirements have to be determined by the method of VAN HOUTUM, ZIJM (regardless of $b(i) = 1$ or $b(i) = 2$). If the encoded solution includes a sequence of several stockpoints with an indicator value of two, cumulative shortfalls have to be determined as shown in 4.1.1.2.

With the inclusion of the additional CLARK, SCARF operation alternative, the number of neighboring solutions is doubled since two, instead of one, neighbors can be generated by a modification of the corresponding indicator variable. If it is impossible to operate with no delays for some stockpoints, the respective encoding $b(i) = 1$ is forbidden. By implementing the presented local search procedures together with some meta strategies like Descent, Simulated Annealing, Tabu Search, or Genetic Algorithms, different operation configurations are evaluated with respect to their cost impact and a reasonably good solution will draw benefits from both presented materials flow characteristics.

4.1.4 Extensions of the Simpson Model

Some assumptions of the presented formulation of the SIMPSON model might appear restrictive for practical applications. In the following, some of these limitations are relaxed and necessary model extensions are discussed together with resulting adaptions of algorithms. Extensions concern the restriction to α-type service measures, the pure make-to-stock assumption, the limitation to deterministic processing times, and lot-for-lot replenishment.

4.1.4.1 Different Service Measure.
The fact that an emergency operation has to be setup may be important for some stockpoints whereas for other installations, only the quantity to be made available by speeding up outstanding orders is of relevance. Therefore, some stockpoints face internal service levels of the α-type, whereas other stockpoints operate subject to γ-service levels.[28] The extension of the SIMPSON approach to different measures of service is two-fold. The first approach forces all stockpoints of a serial system to guarantee a stockout size related γ-service measure, whereas, in a second further extension, some stockpoints operate under α- and others under γ-service level constraints.

Operation with respect to the γ-type service measure is complicated by the negative safety stock problem, that is by the fact that τ_i periods can be covered without holding any safety stock. In the following, the solution will be restricted to non-negative planned safety stocks as discussed in Section 2.3. First, assume that all stockpoints $i = 1, ..., n$ are subject to a predetermined γ-service level constraint. Then, every stockpoint is able to cover $\tau_i = 2\pi\frac{\mu_i}{\sigma_i}(1 - \gamma_i)$ periods demand variability without requiring any safety stock. If service levels γ_i are sufficiently large, that is that $\tau_i \leq \lambda_i$, this will

---

[28] See SCHNEIDER, H., RINKS, D.B., KELLE, P. [1995], where the depot operates under an α-service level constraint and the retailers satisfy γ-service level restrictions in a two-stage divergent system.

only affect the amount of time periods that have to be covered by holding a positive safety stock. If $\tau_i > \lambda_i$ for some stockpoints i, the single processing time of i does not necessitate safety stocks. Additionally, the excess coverage potential $\tau_i - \lambda_i$ is available to cover demand uncertainties during processing times of upstream stockpoints. This can yield to a problem reduction property if cumulative τ-values from the first to the rth stockpoint exceed the corresponding cumulative processing times for at least one stockpoint. In this case, no safety stock is needed for this part of the serial network and the entire system can be optimized without this part.

Property 4.1.1. If a stockpoint $1 \leq r \leq n$ with

$$\sum_{i=1}^{r}(\tau_i - \lambda_i) \geq 0 \qquad \text{and} \qquad (4.10)$$

$$\sum_{i=1}^{q}(\tau_i - \lambda_i) < 0 \qquad \forall q < r \qquad (4.11)$$

exists, stockpoints 1,...,r need no safety stock ($SS_i = 0 \ \forall i = 1, ..., r$) and can be excluded from further analysis.[29]

This reduction property is sequentially applied to the remaining subsystem until no further problem reduction is possible. Note that condition (4.11) can always be fulfilled if the reduction property is applied starting with the most upstream stockpoint. If the serial system under consideration cannot be reduced any further, the coverage time of any stockpoint will never be smaller than the time τ implying zero safety stock.[30] Therefore, the following constraint can be added to the original set of constraints without excluding any optimal solution candidate.

$$T_i \geq \tau_i \qquad i = 1, ..., n. \qquad (4.12)$$

Over the modified (reduced) solution set, the objective function is concave.[31] Minimization of a concave objective function under linear constraints leads to a similar extreme point property as shown for α-service levels. The final-stage stockpoint covers its replenishment lead time plus the review period. The extreme point property implies that stockpoints $i = 1, ..., n - 1$ either cover the time τ_i connected with a safety stock of zero or its replenishment lead time given by the cumulative uncovered upstream processing times less excess downstream coverage potential.

Let $\Delta(i)$ denote excess downstream coverage potential without requiring any safety stock.

---

[29] For a proof see Appendix B, Proposition B.1.1.
[30] For a proof see Appendix B, Proposition B.1.2.
[31] See Appendix A, Lemma A.1.2, A.1.4.

$$\Delta(i) = \max_{i+1 \leq v \leq n} \left\{ \sum_{j=i+1}^{v} (\tau_j - \lambda_j) \right\}^{+}$$

$$= (\Delta(i+1) + \tau_{i+1} - \lambda_{i+1})^{+}$$

for $i = 1, ..., n-1$ and $\Delta(n) = 0$. Then, the evaluation of the extreme point property results in the following relevant coverage values for an optimal policy.[32]

$$T_n^* = \lambda_n + \sum_{j=1}^{n-1} (\lambda_j - T_j^*) \qquad \text{and}$$

$$T_i^* \in \left\{ \tau_i \; ; \; \lambda_i + \sum_{j=1}^{i-1} (\lambda_j - T_j^*) - \Delta(i) \right\} \qquad i = 1, ..., n-1 \quad (4.13)$$

With respect to net coverage requirements $\lambda_i - \tau_i$ this can be interpreted as an "all or nothing"-policy. The safety stock is either zero or it covers the replenishment lead time except for adjustments of safety stock sizes that are connected with cost savings.

An equivalent backward characterization of relevant coverage times for a non-reducible system can be derived. Starting with the most downstream stockpoint, an optimal coverage time for installation i can be represented by solely using the optimal values of downstream stockpoints.

$$T_i^* = \tau_i \quad \text{if} \quad \sum_{j=i+1}^{n} (T_j^* - \lambda_j) > \Delta(i) \quad i = 1, ..., n-1$$

$$T_i^* \in \left\{ \Delta(l-1) + \sum_{j=l}^{i} (\lambda_j - \tau_j) - \Delta(i), \; l = 1, ..., i \right\}$$

$$\text{if} \quad \sum_{j=i+1}^{n} (T_j^* - \lambda_j) = \Delta(i) \quad i = 1, ..., n-1.$$

Under this characterization, the coverage time depends on cumulative excess coverage times of all downstream stockpoints. Only if excess downstream coverage includes the excess that can be created without cost, a positive safety stock is generated by a positive coverage time to deal with cumulative processing times of upstream installations (less τ-coverage). If the excess downstream coverage exceeds $\Delta(i)$, processing time of i has already been covered and the resulting coverage time is τ_i.

In a non-recursive characterization of coverage times, the relevant values can be expressed by

---

[32] Appendix B, Theorem B.1.1.

$$T_n^* \in \left\{ \lambda_n + \Delta(l-1) + \sum_{j=l}^{n-1}(\lambda_j - \tau_j), \ l = 1, ..., n \right\} \qquad \text{and}$$

$$T_i^* \in \{\tau_i\} \cup \left\{ \lambda_i + \Delta(l-1) + \sum_{j=l}^{i-1}(\lambda_j - \tau_j) - \Delta(i), \ l = 1, ..., i \right\}$$
$$i = 1, ..., n-1$$

with $\Delta(0) = 0$. Note that setting $\tau_i = 0$ for all stockpoints yields the same coverage time values as presented for the α-service level case.

The above properties can be utilized to find the optimal safety stock policy under γ-service level constraints by the same methods used for solving the α-service level model. Only a few adjustments are necessary, mainly caused by the τ values. The following notation assumes, that a non-reducible problem is analyzed, and therefore all present problem reduction opportunities have been exploited beforehand.

Adjusted Dynamic Programming Algorithm. Both forward and backward recursion algorithm can alternatively be used in the γ-service level case. Adjustments require the use of the respective safety factor[33] $k_i(u_i)$ for the determination of cost values $c_i(u_i)$, and to adjust state and decision spaces with respect to different extreme points.

- Backward recursion

$$Z_i = \left\{ \lambda_i + \Delta(l-1) + \sum_{j=l}^{i-1}(\lambda_j - \tau_j), \ l = 1, ..., i \right\} \qquad i = 1, ..., n.$$

$U_n(z_n) = \{z_n\}$, and $U_i(z_i) = \{\tau_i \ ; \ z_i - \Delta(i)\}$ for $i = 1, ..., n-1$

- Forward recursion

$$Z_n = \{0\} \qquad \text{and for } i = 1, ..., n-1$$

$$Z_i = \left\{ \Delta(i) \} \cup \{\max\{\Delta(i) \ ; \ \Delta(l-1) + \sum_{j=l}^{i}(\lambda_j - \tau_j)\}, \ l = 1, ..., i \right\}$$

$$U_i(z_i) = \begin{cases} \{\max\{\tau_i \ ; \ \lambda_i + \Delta(l-1) + \sum_{j=l}^{i-1}(\lambda_j - \tau_j) - z_i\}, l = 1, ..., i\} \\ \qquad\qquad \text{if } z_i = \Delta(i) \\ \\ \tau_i \qquad\qquad\qquad\qquad\qquad\qquad\qquad \text{if } z_i > \Delta(i) \end{cases}$$

---

[33] See Section 2.3.1, p. 39.

Adjusted Heuristics. The similarity concerning the all or nothing coverage strategy allows to use the binary representation for a solution. Adjustments are required for the derivation of coverage times that result from a solution representation vector $\underline{b}$.

$$
T_i = \begin{cases}
\tau_i & b(i) = 0 \\
\Delta(p(i)) + \displaystyle\sum_{j=p(i)+1}^{i} (\lambda_j - \tau_j) - \Delta(i) & b(i) = 1
\end{cases}
$$

Using this representation, the solution construction techniques end-item buffering, first- and final-stage buffering, and all-item buffering can be evaluated with respect to resulting coverage times and costs. Cost criterion based successive assignment methods can be implemented in the same way outlined for α-service levels, only simple adjustments with respect to τ-values are necessary. Given an initial solution, neighboring solutions can be generated by switching bits and the solution evaluation follows the above equation.

The use of a different service measure definition will, in general, yield other safety stock size requirements and can additionally provide a different allocation of safety stocks in a multi-echelon system. The approaches presented so far are developed for situations where the same type of service level applies for all stockpoints of the system. The unifying extension where some stockpoints satisfy α- and the remaining stockpoints are subject to γ-service levels constraints is straightforward to the γ-service level approach. Depending on the service measure required for stockpoint i, the following setting has to be implemented.

- Non-stockout probability α_i
  Set $\tau_i = 0$, $k_i = \Phi^{-1}(\alpha_i)$
- Backlog size service level γ_i
  Set $\tau_i = 2\pi \frac{\mu_i}{\sigma_i}(1 - \gamma_i)$, $k_i = k_i(T_i)$

With these settings, the relevant coverage times and algorithms can be applied as outlined for the pure γ-service level model.

4.1.4.2 Positive Customer Service Time.
The SIMPSON and the CLARK, SCARF model assume that the system operates under a make-to-stock regime where immediate response to customer requests is necessary at predetermined service levels, and therefore $S_n = 0$ is required. Only large demand situations allow for backordering of requirements. In practice, a positive customer service time can be observed and is accepted by customers. This depends on product market competition. Nevertheless, the maximum acceptable customer service time will, in general, not be large enough to operate the system in response to orders. Assuming a positive service time bridges the pure make-to-stock environment towards the make-to-order situation. As a consequence, the time span of uncertain demand periods which has to be covered

by holding safety stocks is reduced by the customer service time. The extreme make-to-order case is reached if the service time equals the cumulative processing time,

$$ST_n = \sum_{i=1}^{n} \lambda_i.$$

In this case no safety stock coverage against uncertain demands is necessary. Customer demands can be processed and materials replenished with order acceptance. The only need for safety stocks in such a situation can result from stochastic processing times or scarce capacity.

The solution procedure for $S_n > 0$ can be dealt with by applying the γ-service level logic with the following adjustments.

- α-service levels
  $\tau_n = S_n$, $\tau_i = 0$ $i = 1, ..., n-1$, $k_i = \Phi^{-1}(\alpha_i)$
- γ-service levels
  $\tau_n = S_n + 2\pi\frac{\mu_n}{\sigma_n}(1 - \gamma_n)$, $\tau_i = 2\pi\frac{\mu_i}{\sigma_i}(1 - \gamma_i)$ $i = 1, ..., n-1$, $k_i = k_i(T_i)$

4.1.4.3 Stochastic Processing Times. A very important extension concerns the relaxation of the assumption of deterministic processing times. Besides stochastics in replenishment lead times that result from internal stock insufficiencies, various sources cause processing time variability and uncertainty.[34]

- Machine breakdowns and random work station service
- Capacity constraints and non-controllable/observable job priorities at bottleneck workstations
- Processing times that depend on manufacturing quantities

Literature on modeling serial systems with random lead times under full delay material flow is reviewed and applied by using queuing methodology by DIKS, VAN DER HEIJDEN.[35] The scope of this section is to extend the SIMPSON model to the situation where processing times are random. INDERFURTH[36] distinguishes two planning approaches. Observed from MRP practice, the necessity for deterministic calculations requires planning lead/processing times. Therefore, the stochastic variable is converted into a maximum reasonable time by adding a safety component to the expected value. The second method determines the joint standard deviation of lead time demand as a basis for safety stock planning.[37] The model extension presented by INDERFURTH utilizes the first method for non-final-stage stockpoints and the second approach

---

[34] The processing time definition includes the time span between materials and product availability, whereas the lead time additionally includes the time span between order release and materials availability.

[35] DIKS, E.B., VAN DER HEIJDEN, M.C. [1996].

[36] INDERFURTH, K. [1993].

[37] See Section 2.2.2.2.

for final-stage installations. The following additional assumptions are incorporated.

- Processing times are assumed to be normally distributed $\tilde{\lambda}_i \sim N(\lambda_i, \sigma_{\lambda_i}^2)$.
- Processing time variability is classified into reasonable and extraordinary realizations. Regular variations are covered by safety stocks whereas excess realizations are dealt with by operating flexibility.
- The maximum reasonable processing time level is characterized by the percentile α_{λ_i} which denotes the probability that the processing time realization will not exceed the planned processing time $\hat{\lambda}_i$. Then, the planned processing time for process i in order to guarantee that the fraction α_{λ_i} of replenishment orders arrives within $\hat{\lambda}_i$ time periods is

$$\hat{\lambda}_i = \lambda_i + k_{\lambda_i} \cdot \sigma_{\lambda_i},$$

with $k_{\lambda_i} = \Phi^{-1}(\alpha_{\lambda_i})$.

The approach of INDERFURTH for a serial system with stochastic demand and processing times results in the following optimization problem.

$$\min\ C = \sum_{i=1}^{n-1} h_i \cdot \sigma_i \cdot k_i \cdot \sqrt{T_i} + h_n \cdot k_n \cdot \sqrt{\sigma_n^2 \cdot T_n + \mu_n^2 \cdot \sigma_{\lambda_n}^2}$$

$$s.t. \quad \sum_{j=1}^{i} T_j \leq \sum_{j=1}^{i} \hat{\lambda}_j \qquad i = 1, ..., n-1$$

$$\sum_{j=1}^{n} T_j = \sum_{j=1}^{n-1} \hat{\lambda}_j + \lambda_n$$

$$T_i \quad \geq \quad 0 \qquad \qquad i = 1, ..., n.$$

The advantage of this construction is that the extreme point property is not affected. On the other hand, this approach constructs a safety lead time and simultaneously forces the system to cover demand uncertainty during this safety period. This is only an appropriate approach if every order is exactly filled after the maximum reasonable lead time, which is not the case in all situations. Therefore, this approach induces too large safety stocks and does not benefit from joint coverage against both sources of uncertainty. Another critical aspect is that two different types of internal service levels have to be specified for both demand and processing time.

To overcome this criticism, the SIMPSON idea can be interpreted in a more general way. Stochastic processing times induce stochastic service times. If a safety stock coverage time T_i is introduced as in the deterministic variant, lead time L_i and service time dependencies become

$$\tilde{L}_i = \tilde{S}_{i-1} + \tilde{\lambda}_i,$$
$$\tilde{S}_i = \tilde{L}_i - T_i.$$

Because of the random variable nature of processing times, coverage times are no longer limited by a constant lead time. This problem requires a further assumption that is derived from the single-echelon stochastic lead time model. A safety stock determined by the coverage time multiple of the demand standard deviation covers, at a maximum, demand uncertainty during the replenishment lead time. For a single-echelon consideration this yields the inequality

$$T \cdot \sigma^2 \leq \lambda \cdot \sigma^2 + \mu^2 \cdot \sigma_\lambda^2.$$

Equivalently, the coverage time is restricted by the average processing time that is increased by a safety surplus.

$$T \leq \lambda + \left(\frac{\mu}{\sigma}\right)^2 \cdot \sigma_\lambda^2$$

Increasing processing time variability requires additional coverage weighted by the inverse of the squared coefficient of single period demand variation. This reflects that higher risk diversification effects appear if demand variability is large. When demand variability increases, the directly assigned safety stock $SS = k \cdot \sigma \cdot \sqrt{T}$ increases by the same order (measured by the standard deviation) but the overall coverage requirement given by the inequality decreases which reflects risk diversification in this case.

In contrast to the INDERFURTH approach that distinguishes two different planning situations, this method represents a straightforward extension to the idea of SIMPSON and bridges the two planning situations. Only one service level has to be chosen with respect to quantity flexibility considerations. The upper safety stock limitation exactly represents the single-echelon safety stock that is necessary to cope with joint uncertainty in demand and lead time. Following this line for the entire serial system, cumulative coverage time to the stockpoint under consideration is not allowed to exceed the corresponding sum of average processing times plus the safety increment given by

$$\left(\frac{\mu_i}{\sigma_i}\right)^2 \cdot Var\left(\sum_{j=1}^{i} \tilde{\lambda}_j\right).$$

The analysis of the variance of the sum of random variables differs for independent and dependent processing times.

- Independent processing times

$$Var\left(\sum_{j=1}^{i} \tilde{\lambda}_j\right) = \sum_{j=1}^{i} \sigma_{\lambda_j}^2$$

Individual processing capacities for each process will in general lead to this case.

- Dependent processing times
  Several reasons demand a model that enables the analysis of dependent processing times where individual processing time random variables are correlated with covariance $\sigma_{\lambda_i,\lambda_j}$. Joint processing of different products on the same capacity intuitively induces a positive correlation, whereas, priority of one product over the other will cause negative correlation.

$$Var\left(\sum_{j=1}^{i} \tilde{\lambda}_j\right) = \sum_{j=1}^{i}\sum_{k=1}^{i} \sigma_{\lambda_j,\lambda_k}$$

The incorporation of these assumptions and relationships into the serial safety stock coverage optimization problem yields the following optimization problem formulation.

$$\min \; C= \sum_{i=1}^{n} h_i \cdot \sigma_i \cdot k_i \cdot \sqrt{T_i}$$

$$\text{s.t.} \qquad \sum_{j=1}^{i} T_j \;\le\; \sum_{j=1}^{i} \lambda_j + \left(\frac{\mu_i}{\sigma_i}\right)^2 \cdot Var\left(\sum_{j=1}^{i} \tilde{\lambda}_j\right) \qquad i = 1, ..., n-1$$

$$\sum_{j=1}^{n} T_j \;=\; \sum_{j=1}^{n} \lambda_j + \left(\frac{\mu_n}{\sigma_n}\right)^2 \cdot Var\left(\sum_{j=1}^{n} \tilde{\lambda}_j\right)$$

$$T_i \quad \ge \quad 0 \qquad\qquad\qquad\qquad\qquad i = 1, ..., n.$$

The objective function remains the same for the stochastic processing time case, whereas a constant, non decision variable dependent safety surplus is added to the right hand side of the constraints. Therefore, the extreme point property remains valid though the safety requirements increase. Nevertheless, the same exact and heuristic solution procedures apply with necessary adjustments concerning the addition of a safety processing time surplus.

The integration of some other aspects related to variable processing times is possible along the lines presented in this subsection. Capacity constraints and lot-size dependent processing times can be modeled in an aggregated way by estimating expected value and standard deviation of processing time characteristics from serial queuing networks. Correlated demands with respect to consecutive time periods that were modeled and analyzed for the special cases of an MA 1 process by INDERFURTH[38] can be integrated by utilizing single-echelon considerations outlined in 2.2.2.3.

4.1.4.4 Batch Ordering. Safety stock reducing effects of significant lot-sizes have not been analyzed for multi-echelon systems within the KIMBALL base-stock concept yet. The extension of full delay approaches with respect

---

[38] INDERFURTH, K. [1995].

to setup costs is analyzed by CLARK, SCARF[39]. They show that a simple echelon stock inventory control rule is no longer optimal under the presence of setup costs (except for the case of a setup cost at the most upstream stockpoint). As an approximation, the use of an echelon stock (s, S)-policy is suggested. LAMBRECHT, MUCKSTADT, LUYTEN[40] provide an improved policy parameter determination method. For predetermined lot-sizes in an echelon stock approach, VAN DONSELAAR[41] presents (approximate) closed form square root formulas for the determination of integral safety stocks.

For the determination of safety stocks in batch ordering environments, the hierarchical planning approach suggests to derive strategic safety stock norms, assuming that operational lot-sizes are given. Since a strategic consideration will, in general, only face rough information on the demand level and its variability instead of forecasting the dynamic nature of demands, a predetermination of batch sizes should follow a static multi-stage lot-sizing model. Since multi-stage lot-sizing lies beyond the scope of this work, only the methods of HEINRICH and the review provided by MUCKSTADT, ROUNDY should be mentioned for references to general network directed lot-sizing models.[42]

The incorporation of batch ordering into the SIMPSON model requires the assumption, that operating flexibility with the resulting speeding up opportunities remains available. This implies that complete outstanding lots are speeded up or that lots can be split into sublots in order to provide the required materials. Under this assumption, cycle stocks serve as an additional buffer potential and safety stocks can be reduced. The determination of safety stock requirements at a stockpoint, given its lead time and the predetermined lot-size, follows the respective calculations presented in Section 2.3.2. The coverage time logic that defines the feasible region of the non-linear optimization problem is not affected by batches. Since, especially under batch ordering, part of the replenishment lead time can be covered without any safety stocks, the τ value logic appears again and the approach developed for γ-service level constraints is required, both for α- and γ-service level constraints in the batch ordering case. Because the formulas under α-service level constraints are similar to the lot-for-lot replenishment model under γ-service levels, the extreme point property holds in this case. Under γ-service levels and batch ordering, the analysis of quadratic loss integral expressions with respect to coverage times provides again a concavity property[43] and the coverage time allocation solutions to be considered for finding the optimal policy parameters still apply. The required adjustments concern the determination of the safety stock size where the formulas and algorithms discussed for the single-echelon model apply.

[39] CLARK, A.J., SCARF, H. [1962].
[40] LAMBRECHT, M.R., MUCKSTADT, J.A., LUYTEN, R. [1984].
[41] VAN DONSELAAR, K. [1990].
[42] HEINRICH, C.E., SCHNEEWEISS, C. [1986], HEINRICH, C.E. [1987], MUCKSTADT, J.A., ROUNDY, R.O. [1993].
[43] Appendix A.2.

4.1.4.5 Lost Demand. If internal stock insufficiencies are covered by external instantaneous emergency supplies instead of speeding up outstanding orders from the pipeline, this fraction of internal demand is lost for all upstream stockpoints. The implementation of order-up-to-levels that are derived under the assumption of backordering of internal shortages will overshoot the service level constraints, that is generate excessive safety stocks. Though an order-up-to-policy is not optimal even in the simple single-echelon system, it can be applied as an approximation and only the policy parameters are adjusted with respect to lost demand. Nevertheless, in a multi-echelon system this is complicated by the fact that backpropagation of observed external demand does no longer represent the effective demand if requirements are lost in the downstream part of the supply chain. The effect is that expected demand is lower than expected final products demand and additionally, the standard deviation decreases because large deviations in demand often result in lost internal demands. In order to incorporate lost demands, assume that these effects are only incorporated at the next upper stockpoint. Then, every safety stock can be determined from the single-echelon lost sales model and the multi-echelon system can be coordinated by the SIMPSON model.

4.2 Divergent Systems

4.2.1 Full Delay Approaches

In their paper on optimal policies for serial models, CLARK, SCARF discuss an extension of the echelon stock concept to divergent systems.[44] As in a serial system, internal stock insufficiencies can occur, meaning that a stockpoint has not sufficient stock on hand to satisfy all requests. Additional problems that arise from stock insufficiencies in divergent systems are to allocate the available amount of stock among successors. Besides the determination of an optimal ordering policy with respective parameters, an allocation function has to be determined. This problem turns out to be even more difficult than the ordering policy determination. This is the reason why the overall optimal replenishment strategy for divergent systems is still unknown and can be expected to be rather complex. Nevertheless, restricting the attention to simple ordering policies enables the derivation of nice properties similar to the ones obtained for serial systems. These policies are based on the echelon stock concept. When concentrating on echelon information, that is when the decisions of a stockpoint are based on a single number, allocation of available stock is also based on this single number. Therefore, it may happen that less echelon stock is allocated to one successor than it is already in transit to this installation. This situation would imply that a negative shipment quantity is planned for this successor and is therefore excluded by the

---

[44] CLARK, A.J., SCARF, H. [1960].

so called "balance assumption". DIKS[45] extends the mathematical problem formulation of CLARK, SCARF to the general N-echelon divergent case and discusses the nature of optimal allocation functions. Besides the optimality of an order-up-to-policy with one critical number for each stockpoint, the decomposition principle holds. Therefore, the optimal critical numbers can be computed sequentially. Nevertheless, the problem of determining an allocation rule remains and since the determination turns out to be too complex, relaxations are concerned. FEDERGRUEN, ZIPKIN analyze myopic allocation rules whereas VAN DER HEIJDEN, DIKS, DE KOK compare the performance of different linear allocation functions.[46]

EPPEN, SCHRAGE[47] analyze a two-echelon divergent system with central control policy and derive closed form square root safety stock formulas. The stockpoints at the second stage $i = 1, ..., n$ are all identical with respect to processing times ($\lambda_i = \lambda$) and cost. The single first-stage stockpoint $i = 0$ is not allowed to hold stock and incoming inventory is allocated to the second stage installations by the equal fractile allocation rule where stocks are allocated in a way that each final-stage stockpoint faces the same stockout probability. The total system safety stock requirement derived from their assumptions is

$$SS^{ES} = k \cdot \sqrt{\lambda_0 \cdot \sum_{i=1}^{n} \sigma_i^2 + \lambda \cdot \left(\sum_{i=1}^{n} \sigma_i\right)^2}.$$

EPPEN, SCHRAGE compare this finding with two extreme alternative strategies. The first strategy is the independent operation of all second-stage stockpoints with a cumulative processing time $\lambda_0 + \lambda - 1$ (decentralized system). The total system safety stock for this system is given by

$$SS^D = k \cdot \sqrt{\lambda_0 + \lambda} \cdot \sum_{i=1}^{n} \sigma_i.$$

The second extreme strategy is to operate under a centralized system. All demands are transmitted to a central inventory point and delivered to the customers from this point with a total lead time of $\lambda_0 + \lambda$ time periods. Under this policy, the required safety stock is

$$SS^C = k \cdot \sqrt{(\lambda_0 + \lambda) \cdot \sum_{i=1}^{n} \sigma_i^2}.$$

A comparison of the three systems yields

---

[45] DIKS, E.B. [1997].

[46] FEDERGRUEN, A., ZIPKIN, P. [1984c], VAN DER HEIJDEN, M.C., DIKS, E.B., DE KOK, A.G. [1997].

[47] EPPEN, G., SCHRAGE, L. [1981]

$$SS^D > SS^{ES} > SS^C.$$

Replenishment coordination by the first-stage installation yields two advantages. The depot effect concerns the administrative coordination and bundling of requirements of all final stockpoints which might result in more negotiation power and quantity discounts. The portfolio effect describes the statistical risk pooling effect that central (pooled) demand risk (measured by its variability) is smaller than the sum of individual variations. Consider a two-stage distribution system with one depot and two retailers with a demand correlation coefficient of ρ $(-1 \leq \rho \leq +1)$. Then, the central demand standard deviation is

$$\sigma_C = \sqrt{2 \cdot (1 + \rho)} \cdot \sigma.$$

If demands at both retailers are perfectly positive correlated $(\rho = 1)$, central demand variability equals the sum of both retailers variabilities. In all other cases, central variability is smaller with the extreme of perfect negative correlation $(\rho = -1)$ where central demand becomes deterministic.

Because the explicit safety stock results of EPPEN, SCHRAGE are based on rather limiting assumptions, other approaches for the determination of safety stocks in general divergent networks are required. A numerical method under full delay delivery conditions is provided by DE KOK[48]. Several extensions are available which make this method attractive for practical use. The method assumes that maximum stock levels Δ_i at each non-final-stage stockpoint and linear allocation fractions p_i for all non-first-stage stockpoints are given.[49] For a pure distribution system with no or negligible value added, $\Delta_i = 0$ indicates stockless internal stockpoints. In cases with value added, the optimization of stock levels is a matter of extensions. The outcome of the method are final-stage order-up-to-levels that guarantee predetermined target service levels. Echelon order-up-to-levels for internal stockpoints are computed by adding the respective stock levels Δ_i.

Similar to the serial system algorithm of VAN HOUTUM, ZIJM, the approach computes expressions for stockpoint inventory positions utilizing the shortfall concept. In addition to serial systems, the allocation functions influence the shortfall to a stockpoint. If excess echelon demand exceeds the maximum stock level, the shortage is allocated among the successors by linear fractions p_i and the shortfalls are given by

$$Y_1 = 0$$
$$Y_i = p_i \cdot (Y_{v(i)} + D(\lambda_{v(i)}) - \Delta_{v(i)})^+ \qquad i \in P \cup E.$$

The distribution functions for these expressions are given by incomplete convolutions. Final-stage order-up-to-levels that fulfill target service level con-

[48] DE KOK, A.G. [1990].
[49] VERRIJDT, J.H.C.M., DE KOK, A.G. [1995], [1996].

straints are determined by inserting the shortfall expressions for the final-stage stockpoints into the formulas outlined in 4.1.1.3 and performing $|E|$ independent bisection procedures.

One advantage of this approach is that it is open to several extensions towards practical applicability. One of the major restricting assumptions concerns given maximum stock levels Δ_i. Using the above model in an application, guidelines for the determination of these parameters are required. Within a pure distribution context, this shortcoming is rather unimportant because, when there is no value added, it is reasonable to put all buffer inventory to the final stage, i.e. $\Delta_i = 0 \ \forall i \notin E$. In DE KOK, LAGODIMOS, SEIDEL, this problem is addressed for a two-echelon divergent system by optimizing system inventory holding costs with respect to the variables Δ_i. The extension of this approach to arbitrary divergent systems is nevertheless limited. It turns out that the objective function, even with only one decision variable, is neither convex nor concave and, therefore, the optimization problem has to be solved by cumbersome numerical search techniques.[50] Other extensions concern correlated demands[51], stochastic lead times[52], transshipments[53], and different shipment frequencies[54].

4.2.2 Simpson Modeling Approach

The allocation problem that causes additional complexity in divergent models of the CLARK, SCARF type is not present when the SIMPSON modeling idea is applied. If sufficient operating flexibility is available, the demand of every stockpoint can be supplied by its predecessor within the service time. The serial model can directly be extended to divergent models if a stockpoint guarantees an identical service time to all direct successors. In this situation, the relation of succeeding stockpoints is identical to the serial system.[55]

$$\min \ C = \sum_{i \in A \cup P \cup E} h_i \cdot \sigma_i \cdot k_i \cdot \sqrt{ST_{v(i)} + \lambda_i - ST_i}$$

$$s.t. \quad ST_i \ \leq \ ST_{v(i)} + \lambda_i \qquad \forall i \in A \cup P$$
$$ST_0 \ = \ 0$$
$$ST_i \ = \ 0 \qquad \forall i \in E$$
$$ST_i \ \geq \ 0 \qquad \forall i \in A \cup P$$

The service time ST_i that results from the coverage time T_i of a stockpoint is given by the service time $ST_{v(i)}$ that the immediate predecessor promises to

---

[50] DE KOK, A.G., LAGODIMOS, A.G., SEIDEL, H.P. [1992].
[51] LAGODIMOS, A.G., DE KOK, A.G., VERRIJDT, J.H.C.M. [1995].
[52] VAN DER HEIJDEN, M.C., DIKS, E.B., DE KOK, A.G. [1999].
[53] DIKS, E.B., DE KOK, A.G. [1996], DIKS, E.B., DE KOK, A.G. [1998].
[54] VAN DER HEIJDEN, M.C. [1997b].
[55] INDERFURTH, K. [1991].

all its successors plus the processing time λ_i. For all final-stage stockpoints, service times are assumed to be zero. The resulting optimization model expressed in service times is of the same structure as the serial model. This concave minimization problem under linear constraints connected with an extreme point property yields a cover all or nothing policy concerning the replenishment lead time.

$$ST_i^* \in \{0; ST_{v(i)}^* + \lambda_i\} \qquad i \in A \cup P$$

The equivalent formulation of the above optimization problem in terms of coverage times is

$$\min \ C = \sum_{i=1}^{n} h_i \cdot \sigma_i \cdot k_i \cdot \sqrt{T_i}$$

$$s.t. \qquad \sum_{j \in V(i)} T_j \ \leq \ \sum_{j \in V(i)} \lambda_j \qquad \forall i \in A \cup P$$

$$\sum_{j \in V(i)} T_j \ = \ \sum_{j \in V(i)} \lambda_j \qquad \forall i \in E$$

$$T_i \qquad \geq \quad 0 \qquad i = 1, ..., n$$

and the evaluation of the respective extreme points yields

$$T_i^* = \lambda_i + \sum_{j \in V(i) \backslash \{i\}} (\lambda_j - T_j^*) \qquad \forall i \in E \quad \text{and}$$

$$T_i^* \in \left\{ 0 \ ; \ \lambda_i + \sum_{j \in V(i) \backslash \{i\}} (\lambda_j - T_j^*) \right\} \qquad \forall i \in A \cup P.$$

Safety stocks at a final-stage installation cover the corresponding processing time plus the sum of uncovered upstream processing times. Non-final-stage stockpoints have the additional opportunity of postponing coverage, which implies that their processing times are covered by safety stocks at succeeding stages. Step by step evaluation of these sets leads to the non-recursive characterization

$$T_i^* \in \left\{ \sum_{j \in w(l,i)} \lambda_j \ , \ \forall l \in V(i) \right\} \qquad \forall i \in E \quad \text{and}$$

$$T_i^* \in \{0\} \cup \left\{ \sum_{j \in w(l,i)} \lambda_j \ , \ \forall l \in V(i) \right\} \qquad \forall i \in A \cup P.$$

This extreme point property enables the development of efficient optimization algorithms and heuristics. The Dynamic Programming backward recursion formulation can easily be adapted by exploiting the $1 : |n(i)|$ predecessor-successor relationship of stockpoints in divergent systems. Coverage times are decision variables $(u_i = T_i)$ and replenishment lead times are state variables.

$$z_i = \lambda_i + \sum_{j \in V(i) \setminus \{i\}} (\lambda_j - u_j) \qquad i = 1, ..., n.$$

State transition for all stockpoints $j \in n(i)$ is given by

$$z_j = z_i + \lambda_j - u_i \qquad \forall j \in n(i).$$

For the most upstream stockpoint being externally supplied, $z_1 = \lambda_1$ holds, which means that the external supplier can always deliver. State spaces Z_i and decision spaces $U_i(z_i)$ can explicitly be expressed by exploiting the extreme point properties. Again, only a few values are relevant for the computation of an optimal safety stock allocation policy.

$$Z_i = \left\{ \sum_{j \in w(l,i)} \lambda_j \ , \ \forall l \in V(i) \right\} \qquad i = 1, ..., n.$$

The replenishment lead time for an optimal policy consists of the accumulation of processing times of consecutive stockpoints from stockpoint l to i. The decisions that are relevant for finding an optimal buffer allocation are identical to the replenishment lead time for all final-stage stockpoints. For all non-final stage stockpoints, the optimal coverage decision is either identical to the replenishment lead time or equal to zero.

$$U_i(z_i) = \{z_i\} \quad \forall i \in E \quad \text{and} \quad U_i(z_i) = \{0 \ ; \ z_i\} \quad \forall i \in A \cup P.$$

Costs for final-stage coverage of z_i time periods can be expressed by

$$f_i(z_i) = c_i(z_i) \qquad \forall z_i \in Z_i$$

where $c_i(z_i)$ depends on the objective function criterion under consideration.[56] Functional values for the states of stockpoints $i \in A \cup P$ are derived from

$$f_i(z_i) = \min_{u_i \in U_i(z_i)} \left\{ c_i(u_i) + \sum_{j \in n(i)} f_j(z_i + \lambda_j - u_i) \right\} \forall z_i \in Z_i.$$

The computational complexity of the Dynamic Programming algorithm for divergent systems is affected by the connection of stockpoints. Using a low level coding for stockpoints, the maximum number of states concerning a stockpoint is given by its index. Because of the possibility of multiple successors, the number of states can even be smaller. Therefore, the worst case complexity results from a serial system and is $O(n^2)$. The divergent system

---

[56] See the Dynamic Programming formulation for a serial system.

having the lowest complexity is the two-stage system with a complexity of $O(n)$.

The heuristic solution procedures developed for a serial system apply in the same way for a divergent system. Each final-stage stockpoint is encoded with $b(i) = 1$ and the all or nothing coverage alternatives for first and intermediate stage stockpoints are represented by $b(i) \in \{0, 1\}$. Within successive assignment solution construction and local search solution improvement, the minor adjustment that is required for divergent systems is to compare coverage at a stockpoint with the complementary increment or decrement at all succeeding stockholding installations.

4.2.3 Comparison and Synthesis

The difference between full and no delay approaches becomes more critical in divergent systems. The main advantage of upstream safety stocks besides lower holding costs is provided by the risk pooling effect, that is that the demand standard deviation at first and intermediate stage stockpoints is lower than the sum of the corresponding final-stage demand standard deviations. Therefore, implicitly determined optimal internal service levels will even be lower compared to the serial results. If the no delay approach is confronted with large internal service level constraints induced by limited operating flexibility, an allocation of safety stocks that profits from low holding costs and low standard deviations is made unattractive by excessively large safety factors. Therefore, especially in situations with high internal service level constraints for some stockpoints, a combination of both approaches becomes attractive.

If safety stock levels obtained under the no delay materials flow assumption are implemented in full delay systems, service level deviations can be compensated by a similar adjustment procedure as outlined for serial supply chains. Additionally, allocation fractions are required. Then, the service level achieved by SIMPSON model based safety stocks can be determined by the DE KOK model and the required increments to satisfy service level constraints can be derived by a bisection procedure. These incremental adjustments can be integrated into the solution algorithms along the lines presented for serial systems.

4.2.4 Extensions

Two extensions are presented in this section. The first extension concerns the use of backlog size related service measures for all stockpoints. The second extension is specific for divergent network types with several successors and allows for different service times to successors.

4.2.4.1 Different Service Measure. If each stockpoint operates under a service level of the γ-type, a time span of τ_i periods can be covered by i

with zero safety stock.[57] If $\tau_i \leq \lambda_i \; \forall i$, no additional problems occur and the addition of constraints $T_i \geq \tau_i$ yield the relevant extreme points. The resulting coverage strategy becomes to cover either τ_i or the replenishment lead time. In cases where $\tau_i > \lambda_i$ occurs, the excess coverage potential can be used to cover demand uncertainty during upstream processing time periods. Nevertheless, the extension for a divergent network is not straightforward compared to the results presented for serial network topologies. Additional problems arise from the fact that $T_i \geq \tau_i \; \forall i$ will not hold for the optimal policy of a divergent system in general. An additional constraint $T_i \geq \tau_i$ for a stockpoint with $\tau_i > \lambda_i$ causes a reduction of coverage times for at least one preceding stockpoint $j \in V(i) \backslash \{i\}$. Therefore, downstream stockpoints $k \in N(j)$ on every successor path of j have to increase their coverage times. If these downstream stockpoints face sufficiently large inventory holding costs, this policy causes larger costs than the policy that does not include the additional constraints. Therefore, these constraints would exclude candidates for an optimal policy.

Because of the non-uniqueness of successors, a stockpoint r with $\sum_{j \in V(r)} (\tau_j - \lambda_j) \geq 0$ will not necessarily lead to zero safety stocks for $j \in V(r)$. On the other hand, such a property allows for a decomposition of the problem.

Property 4.2.1. If a stockpoint r with

$$\sum_{j \in V(r)} (\tau_j - \lambda_j) \geq 0 \qquad and \tag{4.14}$$

$$\sum_{j \in V(q)} (\tau_j - \lambda_j) < 0 \qquad \forall q \in V(r) \backslash \{r\} \tag{4.15}$$

exists, then for an optimal policy $SS_r = 0$ holds. If $r \notin E$, the problem can be decomposed into $|n(r)| + 1$ subproblems of divergent structure where every direct successor of r is the root of a new divergent system and the original system is optimized without all stockpoints $\in N(r)$.

Condition (4.15) can always be satisfied if Property 4.2.1 is applied to the system starting with the most upstream stockpoint and until no further decomposition is possible. If the property applies to every path from the root of the original system to all end stockpoints, (a stockpoint r exists on every path $w(1, e) \; \forall e \in E$) decomposition yields a subsystem that includes at least the root of the original system and requires no safety stocks.

Although the coverage time will not be larger or equal to τ_i in general, the property holds that an optimal coverage time will never be smaller than the minimum of τ_i and the replenishment lead time.

---

[57] See the single-echelon model and Section 4.1.4.1 for the γ-service level extension in serial systems.

Property 4.2.2. For a non-decomposable divergent problem under γ-service level constraints, optimal coverage times satisfy

$$T_i^* \geq \min\left\{\tau_i \; ; \; \lambda_i + \sum_{j \in V(i) \backslash \{i\}} (\lambda_j - T_j^*)\right\} \qquad i = 1, ..., n. \qquad (4.16)$$

Proof: Appendix B, Proposition B.2.2.

In case of $\tau_i \leq \lambda_i \; \forall i$, the property is identical to $T_i \geq \tau_i$. The objective function is concave over the solution set with additional constraints $T_i \geq \tau_i$. This is no longer valid if stockpoints with $\tau_i > \lambda_i$ exist. The minimum in (4.16) will depend on all upstream coverage times. To derive relevant values for optimal coverage times, the set of feasible solutions is divided into convex subsets in such a way that, over each subset, the objective function is concave. The optimal policy then results from the extreme points of these subsets.[58]

$$T_i^* = \lambda_i + \sum_{j \in V(i) \backslash \{i\}} (\lambda_j - T_j^*) \qquad \forall i \in E \qquad \text{and}$$

$$T_i^* \in LV_i \cup UV_i \qquad \forall i \in A \cup P$$

where LV_i and UV_i are defined by

$$LV_i = \min\{\tau_i \; ; \; \lambda_i + \sum_{j \in V(i) \backslash \{i\}} (\lambda_j - T_j^*)\} \qquad \text{and}$$

$$UV_i = \{T_i \geq LV_i \mid T_i = \lambda_i + \sum_{j \in V(i) \backslash \{i\}} (\lambda_j - T_j^*) - \Delta(i, e) \; , \; \forall e \in E(i)\}$$

$$\text{with } \Delta(i, e) = \max_{v \in w(i,e) \backslash \{i\}} \{\sum_{j \in w(i,v) \backslash \{i\}} (\tau_j - \lambda_j)\}^+.$$

The coverage time of stockpoint i can obtain a minimum value that corresponds to zero safety stocks. A positive safety stock covers the processing time of the related stockpoint plus the sum of uncovered processing times of upstream stockpoints (replenishment lead time) minus an excess downstream coverage potential $\Delta(i, e)$. This coverage potential denotes the time that can be covered by stockpoints on the path from stockpoint i to the end item stockpoint e (i excluded) in excess of their processing times without holding safety stocks. Since usage of a coverage potential of one path affects the other paths with additional costs, every single path from the stockpoint under consideration to the final stage with a positive excess coverage potential has to be taken into account.

Applying optimal coverage time values for the special case of a non-reducible serial system, the lower value LV_i is always equal to τ_i. Otherwise,

---

[58] For a proof see Appendix A.3.

a further application of problem reduction property 4.2 is possible. $\Delta(i, e)$ reduces to $\Delta(i)$ for a unique final product. Therefore, UV_i is equivalent to the replenishment lead time value less $\Delta(i)$ for a serial system. For the non-recursive numerical characterization of optimal coverage time values, we have to ensure that cumulative values of processing times minus zero safety stock coverage times on a path from i to j are not negative. Therefore, values

$$L_{i,j,e} = \begin{cases} [\lambda_j - \tau_j + L_{i,v(i),e}]^+ & \text{if } j \in N(i) \\ \Delta(v(i), e) & \text{if } j \in v(i), \end{cases} \qquad \text{with } \Delta(v(1), e) = 0 \ \forall e \in E$$

are (recursively) defined that take into account an excess coverage potential at stockpoint $v(i)$ with respect to a final-stage stockpoint $e \in E(v(i))$. Then, optimal coverage times can be characterized by

$$T_i^* \in \{\lambda_i + L_{l,v(i),e}, \ \forall l \in V(i), \ \forall e \in E(l)\} \qquad \forall i \in E \qquad \text{and}$$

$$T_i^* \in \{\tau_i\} \cup \{[\lambda_i + L_{l,v(i),\tilde{e}} - \Delta(i, e)]^+, \forall l \in V(i), \ \forall e \in E(i), \ \forall \tilde{e} \in E(l)\}$$
$$\forall i \in A \cup P.$$

The solution algorithms presented for α-service level constraints have to be adjusted to the different optimality property in the γ-service level case.

Adjusted Dynamic Programming Algorithm. Decision and state spaces $U_i(z_i)$, Z_i for a backward recursion algorithm have to be adjusted in the following way.

$$Z_i = \{\lambda_i + L_{l,i,e} \ , \ \forall l \in V(i) \setminus \{i\}, \ \forall e \in E(l)\} \qquad i = 1, ..., n$$

$$U_i(z_i) = \{z_i\} \ \forall i \in E \quad \text{and}$$

$$U_i(z_i) = \begin{cases} \{\tau_i\} \cup \{\max\{\tau_i \ ; \ z_i - \Delta(i, e)\}, \ \forall e \in E(i)\} & \text{if } z_i > \tau_i \\ \{z_i\} & \text{if } z_i \le \tau_i \end{cases}$$
$$\forall i \in A \cup P.$$

For a divergent system, the computational complexity of the Dynamic Programming algorithm is different for γ-service levels. The same complexity as under α-service level only holds if $\tau_i \le \lambda_i \ \forall i$. The Reductions $\Delta(i, e)$ have to be taken into account and result in additional states and in more than two possible decisions for every stockpoint. Let $f = |E|$ denote the number of final-stage stockpoints. In the worst case, on every path to an end item a reduction is possible. This results in f states with respect to reductions of every direct predecessor. Additionally, in every state there might be at most f alternative decisions. Therefore, an upper bound for complexity is $O((fn)^2)$.

Adjusted Heuristics. Additional solution alternatives with respect to successor path reductions also influence the heuristic solution procedures. Nevertheless, as a starting point it can be assumed that reduction effects will only be small and can, therefore, be neglected. Then, similar procedures discussed for the serial γ-service level extension apply with a simple binary encoding of coverage allocation solutions.

In order to incorporate the relevant reduction alternatives, an extended encoding scheme is required. Let $b(i) = 0$ indicate that τ_i periods are covered. The other possible values of $b(i)$ are the indices $j \in E(i)$ of all final-stage stockpoints to i where an excess of τ values can stem from. Then, the relevant coverage time is given by

$$T_i = \max\{\tau_i, \lambda_i + L_{p(i),i,b(p(i))} - \Delta(i, b(i))\}$$

where $p(i)$ denotes the next predecessor to i with $b(p(i)) > p(i)$.

4.2.4.2 Service Time Differentiation. The operating characteristic with identical service times ST_i for all direct successors represents a straightforward extension of the SIMPSON model formulation. Nevertheless, in a divergent multi-echelon system with several direct successors, the implementation of identical service times to all successors $j \in n(i)$ is to be regarded as an additional model assumption and its relaxation might result in a better cost performance. When a stockpoint faces non-identical successors with different inventory holding costs, different processing times, or different service level requirements, the implementation of different service times to successors can result in cost improvements. This relaxation of model assumptions additionally relates to the aspect of customer or successor priorities.

Service times being defined as the time span after which requested material is delivered to successor stockpoints can be implemented in different ways. The natural extension is to let every stockpoint i quote different service times $ST_{i,j}$ to each of its successors $j \in n(i)$. Then, internal demands for material of stockpoint i by successor j in period t are delivered in $t + ST_{i,j}$. As an example, a successor or a customer with high priority may be supplied immediately ($ST_{i,j} = 0$) whereas material requirements for uncritical stockpoints may be delivered after the replenishment lead time ($ST_{i,j} = ST_{v(i),i} + \lambda_i$). This operating policy is analyzed in Model 1. However, the idea behind different service times can be interpreted in a more general way. If a stockpoint i faces a replenishment lead time generated by the predecessor service time $ST_{v(i),i}$ plus the processing time λ_i and implements different service times $ST_{i,j}$ to its successors j for reasons of successor priority connected with better cost performance, the stockpoint might also accept different service times from its predecessor. In this case predecessors already quote different service times for material requirements of downstream stockpoints. By expanding this idea in the most extensive way, all internal demands are driven by customer requests at the final-stage stockpoints $e \in E$. Therefore, every stockpoint implements a service time $ST_{i,e}$ directed to the final-stage requirements of $e \in E(i)$. This control mechanism is analyzed in Model 2.

As a consequence of differentiating service times among several successors, the coverage time decision variable has to be interpreted in an extended way. Different service times announced to successors have to be analyzed with respect to their impact on safety stock coverage requirements. Given the set of service times guaranteed to immediate successor requirements (Model 1) or to final-stage stockpoint requirements (Model 2) respectively, uncertain demands over a span of $T_{i,j}$ or $T_{i,e}$ time periods have to be covered by holding safety stocks. Then, the relationship between service times and safety stock coverage times can be stated for the two different models.

- Model 1

  In this modeling approach where each stockpoint $i \in A \cup P$ supplies its successors $j \in n(i)$ after an individual service time $ST_{i,j}$, the safety stock coverage time $T_{i,j}$ of i with respect to successor j is given by the service time $ST_{v(i),i}$ of the immediate predecessor plus processing time λ_i minus service time $ST_{i,j}$ proposed to successor j.

$$T_{i,j} = ST_{v(i),i} + \lambda_i - ST_{i,j} \;\; \forall i \in A \cup P, j \in n(i)$$
$$T_{e,e} \;\;= ST_{v(e),e} + \lambda_e \;\;\;\; \forall e \in E$$

- Model 2

  When operation is organized in direct response to final-stage requirements where products demanded by stockpoint e are forwarded to the single successor on the path $w(i,e)$ after the service time $ST_{i,e}$, the safety stock coverage time requirement $T_{i,e}$ at i with respect to final product e is given by the service time $ST_{v(i),e}$ guaranteed by the immediate predecessor for final stockpoint e plus processing time λ_i minus service time $ST_{i,e}$.

$$T_{i,e} = ST_{v(i),e} + \lambda_i - ST_{i,e} \;\; \forall i \in A \cup P, e \in E(i)$$
$$T_{e,e} = \;\;\;\; ST_{v(e),e} + \lambda_e \;\;\;\; \forall e \in E$$

The SIMPSON model suggests to connect safety stocks SS_i to the standard deviation of demand during the coverage time. For the service time differentiation extension, there is no explicit equivalence between the safety stock and a single safety stock coverage time variable. Therefore, it is necessary to find the standard deviation of the random variable that represents aggregate demand over the time periods to be covered by stockpoint i in order to satisfy the corresponding service times. It is obvious that the standard deviation of this aggregate is highly influenced by the cross product demand correlation. Because analysis and resulting algorithms can be simplified in case of completely uncorrelated final-product demands, correlated and uncorrelated demands are analyzed separately, though the second is only a special case of the first. When final product demands are completely uncorrelated, the standard deviation of aggregate demand of an internal stockpoint $i \in A \cup P$ generated by the vector $\underline{T}$ of coverage times is given by:

- Model 1

$$\sigma_i^2(\underline{T}) = \sum_{j \in n(i)} \sum_{e \in E(j)} a_{i,e}^2 \cdot \sigma_e^2 \cdot T_{i,j},$$

- Model 2

$$\sigma_i^2(\underline{T}) = \sum_{e \in E(i)} a_{i,e}^2 \cdot \sigma_e^2 \cdot T_{i,e}.$$

Coverage times for different successors or final products can refer to demands of the same time periods. With demands being correlated across products, covariance expressions for coverage times concerning overlapping periods must be taken into account. Consider stockpoint i with a pair (a, b) of successors (Model 1, $a, b \in n(i)$) or final-stage stockpoints (Model 2, $a, b \in E(i)$). The service times of predecessor $v(i)$ with respect to a, b are given by $ST_{v(i),a}, ST_{v(i),b}$ respectively. The coverage times $T_{i,a}, T_{i,b}$ generate service times $ST_{i,a}, ST_{i,b}$. The relationship between these variables is given by

$$T_{i,a} = ST_{v(i),a} + \lambda_i - ST_{i,a},$$
$$T_{i,b} = ST_{v(i),b} + \lambda_i - ST_{i,b}.$$

Coverage time $T_{i,a}$ covers demand uncertainty during the time interval $[t + ST_{i,a}, ..., t + ST_{v(i),a} + \lambda_i]$, whereas $T_{i,b}$ refers to $[t + ST_{i,b}, ..., t + ST_{v(i),b} + \lambda_i]$. Therefore, the overlapping time interval is

$$[\max\{ST_{i,a}, ST_{i,b}\}, ..., \min\{ST_{v(i),a}, ST_{v(i),b}\} + \lambda_i].$$

The two cases where no overlapping occurs are given by

- $ST_{i,a} \geq ST_{v(i),b} + \lambda_i \Leftrightarrow T_{i,a} \leq ST_{v(i),a} - ST_{v(i),b}$
- $ST_{i,b} \geq ST_{v(i),a} + \lambda_i \Leftrightarrow T_{i,b} \leq ST_{v(i),b} - ST_{v(i),a}.$

For Model 1, $ST_{v(i),a} = ST_{v(i),b} = ST_{v(i)}$ holds, the overlapping time interval is never empty, and the number of overlapping time periods is given by

$$\begin{aligned}O_i(a, b) &= ST_{v(i)} + \lambda_i - \max\{ST_{i,a}, ST_{i,b}\} \\ &= ST_{v(i)} + \lambda_i - \max\{ST_{v(i)} + \lambda_i - T_{i,a}, ST_{v(i)} + \lambda_i - T_{i,b}\} \\ &= -\max\{-T_{i,a}, -T_{i,b}\} \\ &= \min\{T_{i,a}, T_{i,b}\}.\end{aligned}$$

In Model 2, the case of no overlapping can occur. Therefore, the number of overlapping time periods is given by (note that the max-operator excludes the cases of a negative number)

$$\begin{aligned}O_i(a, b) &= (\min\{ST_{v(i),a}, ST_{v(i),b}\} + \lambda_i - \max\{ST_{i,a}, ST_{i,b}\})^+ \\ &= (-\max\{(ST_{v(i),a} - ST_{v(i),b})^+ - T_{i,a}, (ST_{v(i),b} - ST_{v(i),a})^+ - T_{i,b}\})^+ \\ &= \max\{0, \min\{T_{i,a} - (ST_{v(i),a} - ST_{v(i),b})^+, T_{i,b} - (ST_{v(i),b} - ST_{v(i),a})^+\}\}.\end{aligned}$$

Taking into account the number of overlapping time periods and the corresponding demand correlation, the standard deviation of the demand aggregate at stockpoint i is given by

- Model 1

$$\sigma_i^2(\underline{T}) = \sum_{j \in n(i)} \sum_{k \in n(i)} \sum_{e \in E(j)} \sum_{f \in E(k)} a_{i,e} a_{i,f} \sigma_e \sigma_f \rho_{e,f} O_i(j,k) \; \forall i \in A \cup P,$$

- Model 2

$$\sigma_i^2(\underline{T}, \underline{ST}) = \sum_{j \in E(i)} \sum_{k \in E(i)} a_{ij} a_{ik} \sigma_j \sigma_k \rho_{jk} O_i(j,k) \; \forall i \in A \cup P.$$

Incorporating the service time logic and the relation of safety stocks to aggregate demand standard deviations, the optimization problem minimizes safety stock holding cost resulting from implemented service times. The constraints ensure that the service time after safety stock coverage cannot exceed the service time before coverage plus processing time. For the two modeling approaches, the service time optimization problems are given by:

- Model 1

$$\min \; C = \sum_{i \in A \cup P \cup E} h_i \cdot k_i \cdot \sqrt{\sigma_i^2(\underline{ST})}$$

$$\begin{aligned}
s.t. \quad & ST_{i,j} \leq ST_{v(i),i} + \lambda_i & \forall i \in A \cup P, \; j \in n(i) \\
& ST_{0,1} = 0 \\
& ST_{e,e} = 0 & \forall e \in E \\
& ST_{i,j} \geq 0 & \forall i \in A \cup P, j \in n(i).
\end{aligned}$$

- Model 2

$$\min \; C = \sum_{i \in A \cup P \cup E} h_i \cdot k_i \cdot \sqrt{\sigma_i^2(\underline{ST})}$$

$$\begin{aligned}
s.t. \quad & ST_{i,e} \leq ST_{v(i),e} + \lambda_i & \forall i \in A \cup P, e \in E(i) \\
& ST_{0,e} = 0 & \forall e \in E \\
& ST_{e,e} = 0 & \forall e \in E \\
& ST_{i,e} \geq 0 & \forall i \in A \cup P, e \in E(i).
\end{aligned}$$

By substitution of service times by safety stock coverage times, the equivalent optimization problems are

- Model 1

$$\min\ C = \sum_{i=1}^{n} h_i \cdot k_i \cdot \sqrt{\sigma_i^2(\underline{T})}$$

$$s.t. \qquad \sum_{j \in V(i) \backslash \{1\}} T_{v(j),j} \leq \sum_{j \in V(i)} \lambda_j \qquad \forall i \in A \cup P$$

$$\sum_{j \in V(e) \backslash \{1\}} T_{v(j),j} = \sum_{j \in V(e)} \lambda_j \qquad \forall e \in E$$

$$T_{i,j} \qquad\qquad \geq \quad 0 \qquad\qquad i \in A \cup P,\, j \in n(i).$$

- Model 2

$$\min\ C = \sum_{i=1}^{n} h_i \cdot k_i \cdot \sqrt{\sigma_i^2(\underline{T})}$$

$$s.t. \qquad \sum_{j \in V(i)} T_{j,e} \leq \sum_{j \in V(i)} \lambda_j \qquad \forall i \in A \cup P;\ e \in E(i)$$

$$\sum_{j \in V(e)} T_{j,e} = \sum_{j \in V(e)} \lambda_j \qquad \forall e \in E$$

$$T_{i,e} \qquad\qquad \geq \quad 0 \qquad\qquad i = 1, ..., n;\ e \in E(i).$$

After presenting optimization problem formulations, the question arises if similar extreme point properties can be derived for the service time differentiation extensions. For Model 1 with service time differentiation with respect to immediate successors an extreme point property can directly be derived.

$$ST_{i,j}^* \in \{0, ST_{v(i),i}^* + \lambda_i\} \qquad\qquad \forall i \in A \cup P, j \in n(i)$$

For the equivalent coverage time formulation, the result becomes

$$T_{e,e}^* = \quad \lambda_e + \sum_{j \in V(e) \backslash \{1,e\}} (\lambda_j - T_{v(j),j}^*) \qquad \forall e \in E$$

$$T_{i,e}^* \in \left\{ 0\ ;\ \lambda_i + \sum_{j \in V(i) \backslash \{1,i\}} (\lambda_j - T_{v(j),j}^*) \right\} \qquad i \in A \cup P,\ e \in E(i).$$

Proof: The extreme point property follows from concavity of the objective function. This is a result of the underlying basic concavity properties and transformations that are included in the adjusted expressions for demand variances. The necessary properties are summarized and proven in the book of SYDSAETER, HAMMOND.[59]

(1) The function given by the minimum of linear functions is concave.
(2) $\sigma_i^2()$ as weighted sum of concave functions is a concave function.

[59] SYDSAETER, K., HAMMOND, P.J. [1995], p. 628.

(3) A concave function applied to the argument of a monotonously increasing concave function represents a concave function.   □

The problem of empty overlapping time intervals in Model 2 complicates the analysis considerably. In this case, the max-operator in the expression for the standard deviation causes the general concavity property to vanish. Nevertheless, by applying a similar technique as used for the γ-service level optimization problem, where the solution set is divided into convex subsets over each of them, the objective function is concave, the following values have to be taken into account to find the optimal policy.

$$T^*_{e,e} = \lambda_e + \sum_{j \in V(e) \setminus \{e\}} (\lambda_j - T^*_{j,e}) \qquad e \in E$$

$$T^*_{i,e} \in \left\{ 0 \; ; \; \lambda_i + \sum_{j \in V(i) \setminus \{i\}} (\lambda_j - T^*_{je}) ; (ST_{v(i),e} - ST_{v(i),f})^+ \; \forall e, f \in E(i) \right\}$$
$$i \in A \cup P, \; e \in E(i)$$

Proof: In cases with no overlapping of coverage intervals, the objective function given for the case without demand correlation applies and the concavity is obvious. In cases where the overlapping interval is non-empty, the max-operator is not relevant. Then, the reasoning presented for Model 1 applies. Therefore, over both of these regions the objective is concave. To get the results for the entire solution set, the feasible region has to be divided into the above cases by additional constraints

$$T_{i,a} \geq ST_{v(i),a} - ST_{v(i),b}$$
$$T_{i,b} \geq ST_{v(i),b} - ST_{v(i),a}$$

where only one right hand side is positive. Applying a recursive argumentation scheme, the values $T_{i,e} = (ST_{v(i),e} - ST_{v(i),f})^+ \; \forall e, f \in E(i)$ have to be considered for an optimal policy.   □

In order to find optimal safety stocks that result from these properties, the algorithmic principles discussed in previous sections apply.

Algorithm Model 1. The implementation of service time differentiation requires one-dimensional state and two-dimensional decision variables for each non-final-stage stockpoint. For $i \in E$, both state and decision variables are one-dimensional. States z_i characterize the replenishment lead time of i whereas decisions $u_{i,j}$ denote the safety stock coverage time of i with respect to material requirements of successor stockpoint j. Let $\underline{u_i} = (u_{i,j})_{j \in n(i)}$ denote the vector of decisions of stockpoint i. For the final stage without successors, these variables become u_i. Similar to the undifferentiated divergent model, the state transition equation is given by

$$z_j = z_i + \lambda_j - u_{i,j} \qquad \qquad \forall i \in A \cup P, j \in n(i)$$

with starting condition $z_1 = \lambda_1$. Let Z_i denote the state space and $U_{i,j}(z_i)$ the decision space for component j within the decision vector of i. Then, state and (component) decision spaces are identical to the model with identical service times. The entire decision space is given by

$$U_i(z_i) = U_{i,1(n(i))} \times \cdots \times U_{i,|n(i)|(n(i))}$$

where $j(n(i))$ denotes the jth element of the set $n(i)$. The functional values for final-stage stockpoints are $f_i(z_i) = c_i(z_i) \ \forall i \in E$. For other stockpoints, they are determined from the functional equation which is adjusted with respect to the decision vectors.

$$f_i(z_i) = \min_{\underline{u}_i \in U_i(z_i)} \left\{ c_i((u_{i,j})_{j \in n(i)}) + \sum_{j \in n(i)} f_j(z_i + \lambda_j - u_{i,j}) \right\} \quad \forall i \in A \cup P$$

Different coverage times require a joint evaluation of $2^{|n(i)|}$ decision combinations which increases the computational effort considerably.

Algorithm Model 2. This further extension leads to vector valued state variables. The state variables $z_{i,e}$ denote the replenishment lead time with respect to customer demands for final product e ($i \in A \cup P, e \in E(i)$). Decision variables $u_{i,e}$ represent the safety stock coverage time for requirements resulting from e ($i \in A \cup P, e \in E(i)$). The respective vectors are $\underline{z}_i = (z_{i,e})_{e \in E(i)}$ and $\underline{u}_i = (u_{i,e})_{e \in E(i)}$. For the final stage, $z_{e,e} = z_e$ and $u_{e,e} = u_e$ holds. The adjusted state transition equations are

$$z_{i,e} = z_{v(i),e} + \lambda_i - u_{v(i),e} \quad \forall i \in P, e \in E(i)$$

with starting condition $z_{1,e} = \lambda_1, \forall e \in E$. State and decision spaces for vector components $Z_{i,e}, U_{i,e}(z_{i,e})$ are still defined as in the standard case.

$$Z_i = Z_{i,1(E(i))} \times \cdots \times Z_{i,|E(i)|(E(i))}$$
$$U_i(\underline{z}_i) = U_{i,1(E(i))}(z_{i,1(E(i))}) \times \cdots \times U_{i,|E(i)|(E(i))}(z_{i,|E(i)|(E(i))})$$

Final-stage functional values are

$$f_e(z_e) = c_e(z_e) \ \forall e \in E.$$

Determination of $f_i(\underline{z}_i) \ \forall i \in A \cup P$ necessitates a joint evaluation of $2^{|E(i)|}$ decisions.

$$f_i(\underline{z}_i) = \min_{\underline{u}_i \in U_i(\underline{z}_i)} \left\{ c_i((u_{i,e})_{e \in E(i)}) + \sum_{j \in n(i)} f_j(\underline{z}_j) \right\} \quad \forall i \in A \cup P$$

The three alternative model formulations differ with respect to their cost performance, computational requirements, and their organizational implementability. Since the base model extension is a special case of Model 1 which itself is a special case of Model 2, the rank order with respect to cost performance is: Base model $\prec$ Model 1 $\prec$ Model 2. The required computational effort in order to find the optimal coverage allocation solutions is Base model $\succ$ Model 1 $\succ$ Model 2. Besides cost advantages and computation time disadvantages, the necessary organizational efforts to implement and to operate under a materials coordination concept should not be disregarded. The operation under a single service time for all requirements is much easier than managing each final-product requirements separately, which implies Base model $\succ$ Model 1 $\succ$ Model 2.

Instead of determining differentiated service time policies by a cumbersome optimal Dynamic Programming procedure, binary encoding based heuristics can be developed. The definition of the binary string with respect to replenishment lead time coverage is extended to double indexed bits due to the coverage provided for each successor (Model 1) or each adjoint final product (Model 2). The respective coverage time and solution quality evaluation is similar to the serial model.

4.3 Convergent Systems

4.3.1 Full Delay Approaches

Problems that occur when determining stock norms for convergent systems have been analyzed first by SCHMIDT, NAHMIAS[60] for a two-echelon special case. In two later papers, it was shown that for systems operating under full delay materials flow, a convergent system can be transformed into an equivalent serial problem.[61] For finding optimal stocknorms in a convergent system, the equivalent serial system has to be derived and the analysis of Section 4.1.1 applies. The serial equivalent to an arbitrary convergent system can be found as follows. Without loss of generality, assume that the stockpoints of a convergent system are numbered in the order of decreasing cumulative processing time from the stockpoint to the final-stage installation, i.e.

$$i \leq j \Leftrightarrow \sum_{k \in N(i)} \lambda_k \leq \sum_{k \in N(j)} \lambda_k \qquad \text{for any } i, j \in A \cup P.$$

If the above condition is violated, the following algorithm finds a numbering of stockpoints $i = 1, ..., n$ that meets the condition. Let N denote the set of already numbered stockpoints and W the set of still unnumbered stockpoints.

[60] SCHMIDT, C.P., NAHMIAS, S. [1985].
[61] ROSLING, K. [1989], LANGENHOFF, L.J.G., ZIJM, W.H.M. [1990].

- $W := A \cup P \cup E$, $N := \emptyset$
- assign number n to the stockpoint $i \in E$ without successor, $N := N \cup E$, $W := W \backslash E$
- assign numbers $n - 1$ to 1 to stockpoints without numbers that satisfy
  (a) $n(i) \in N$
  (b) $i := \arg \min_{j \in W} \{CL_j = \sum_{k \in N(j)} \lambda_k\}$
  $N := N \cup \{i\}$, $W := W \backslash \{i\}$
- adjust processing time $\lambda_i := CL_i - CL_{i+1}$

4.3.2 Simpson Modeling Approach

The planning logic of SIMPSON that every stockpoint guarantees a service time to its successor that depends on its replenishment lead time and the safety stock coverage decision has to be adjusted to the presence of multiple immediate predecessors. For this kind of system, the replenishment lead time for starting the processing of an item is determined by the component with the longest service time. Therefore, the replenishment lead time for given service times of all immediate predecessors is given by the maximum predecessor service time plus the processing time of i.

$$L_i = \lambda_i \qquad i \in A$$
$$L_i = \max_{j \in v(i)} \{ST_j\} + \lambda_i \ i \in A \cup P$$

By incorporating this extension into the serial service time optimization problem and adjusting the objective function and the constraints, the following non-linear optimization problem has to be solved to find optimal service times.

$$\min C = \sum_{i \in A} h_i \cdot \sigma_i \cdot k_i \cdot \sqrt{\lambda_i - ST_i}$$
$$+ \sum_{i \in P} h_i \cdot \sigma_i \cdot k_i \cdot \sqrt{\max_{j \in v(i)} \{ST_j\} + \lambda_i - ST_i}$$
$$+ \sum_{i \in E} h_i \cdot \sigma_i \cdot k_i \cdot \sqrt{\max_{j \in v(i)} \{ST_j\} + \lambda_i}$$

$$s.t. \qquad 0 \le ST_i \le \lambda_i \qquad\qquad \forall i \in A$$
$$0 \le ST_i \le \max_{j \in v(i)} \{ST_j\} + \lambda_i \qquad \forall i \in P$$
$$ST_i = 0 \qquad\qquad\qquad \forall i \in E.$$

The set of service times candidates for an optimal policy can be reduced by some properties that are violated by suboptimal service time combinations.

Property 4.3.1. For an optimal service time policy of a convergent system, service times of stockpoints with the same immediate successor are either identical or coverage times of all predecessors to the point with smaller service time are equal to zero.

$$ST_i^* > ST_j^* \Rightarrow T_k^* = 0 \; \forall k \in V(j)$$

Otherwise, service times can be increased by reducing respective coverage times without affecting the evaluation of the maximum-operator. Property 4.3.1 can be transformed into a characterization for the critical path stockpoint $a(i) \in A(i)$.

Property 4.3.2. In the α-service level case, the maximum predecessor service time results from the path with the largest cumulative processing time.

$$a(i) = \arg \max_{a \in A(i)} \{ \sum_{j \in w(a,i)} \lambda_j \}.$$

Exploiting this property[62] and replacing service times by coverage times, the optimization problem depends only on coverage times.

$$\min \; C = \sum_{i=1}^{n} h_i \cdot \sigma_i \cdot k_i \cdot \sqrt{T_i}$$

$$\text{s.t.} \quad \sum_{j \in w(a(i),i)} T_j \leq \sum_{j \in w(a(i),i)} \lambda_j \qquad i = 1, ..., n$$

$$\sum_{j \in N(i)} T_j \geq \sum_{j \in N(i)} \lambda_j \qquad \forall i \in A$$

$$T_i \geq 0 \qquad i = 1, ..., n.$$

The first type of constraints ensures that on the path from $a(i)$ to i not more than the largest cumulative processing time is covered. The second type enforces all cumulative processing times to be covered. The analysis of the optimization problem leads to an extreme point property as seen in the cases of serial and divergent systems. Optimal coverage times T_i^* obtain values that result from extreme points of the solution set.

$$T_i^* \in \left\{ [\sum_{j \in w(l,i)} \lambda_j - \sum_{j \in N(i) \setminus \{i\}} (T_j^* - \lambda_j)]^+ \; , \; \forall l \in V(i) \right\} \quad i = 1, ..., n$$

The safety stock of stockpoint i covers (positive) differences between cumulative processing times of an upstream path from l to stockpoint i and times already covered by safety stocks at downstream stockpoints in excess of their processing times. The set for T_i^* contains more elements than result directly from the extreme points. If one of the downstream coverage times T_j^* has already covered the processing time of an upstream stockpoint i, the coverage time of i will always be zero. Nevertheless, the exact expression would

---

[62] For a proof see Appendix B, Proposition B.3.1.

require to differentiate between downstream decisions and, therefore, require additional state variables in a Dynamic Programming algorithm. Taking this fact into account, the set becomes equivalent to the serial representation.

In a non-recursive definition, relevant values for optimal coverage times for $i = 1, ..., n$ are

$$T_i^* \in \left\{ [\sum_{j \in w(l,i)} \lambda_j - [\sum_{j \in w(r,n)} \lambda_j - \sum_{j \in w(n(i),n)} \lambda_j]^+]^+, \forall l \in V(i), \forall r \in V(n) \right\}.$$

For a convergent system, every safety stock coverage decision influences all preceding stockpoints. Therefore, a Dynamic Programming algorithm to find the optimal coverage allocation to a convergent system can be derived from the serial forward formulation. Decision variables u_i are safety stock coverage times ($u_i = T_i, i = 1, ..., n$). The state of a stockpoint is defined as the excess coverage of the (unique) path from the direct successor to the final-product stockpoint.

$$z_i = \sum_{j \in N(i) \setminus \{i\}} (u_j - \lambda_j) \qquad i = 1, ..., n$$

For the state transition, excess coverage to the successor is increased by the coverage decision and is decreased by the corresponding processing time.

$$z_i = z_{n(i)} - \lambda_{n(i)} + u_{n(i)} \qquad i = 1, ..., n - 1$$

After forward recursion, backtracking starts with state $z_n = 0$ for the most downstream stockpoint. The state space for every stockpoint and the decision space of possible coverage alternatives in a state are

$$Z_n = \{0\} \qquad \text{and}$$

$$Z_i = \left\{ [\sum_{j \in w(r,n)} \lambda_j - \sum_{j \in w(n(i),i)} \lambda_j]^+, \forall r \in V(n) \right\} \quad i = 1, ..., n - 1$$

$$U_i(z_i) = \left\{ [\sum_{j \in w(l,i)} \lambda_j - z_i]^+, \forall l \in V(i) \right\} \qquad i = 1, ..., n.$$

For first-stage stockpoints $a \in A$, a positive coverage time is only necessary if the processing time exceeds the value of excess downstream coverage z_a. Therefore, the functional values are

$$f_a(z_a) = c_a([\lambda_a - z_a]^+).$$

State dependent costs of stockpoints $i \notin A$ result from the functional equation

$$f_i(z_i) = \min_{u_i \in U_i(z_i)} \left\{ c_i(u_i) + \sum_{j \in v(i)} f_j(z_i + u_i - \lambda_i) \right\} \forall z_i \in Z_i.$$

By finding that every stockpoint covers differences of cumulative processing times to predecessors and processing times covered by downstream stockpoints in excess to their processing times, at most $n - 1$ of these differences are positive. Summing over n stockpoints with, at most, $n - 1$ possible decisions in every state, an upper bound for the computational complexity of the algorithm is $O(n^3)$.

4.3.3 Different Service Measure

An operation under γ- instead of α-service levels only requires slight modifications for a convergent system.

Property 4.3.3. For an optimal policy, service times of stockpoints with the same immediate successor are either identical, or coverage times of all predecessors k with smaller service time are equal to τ_k.

$$ST_i^* > ST_j^* \Rightarrow T_k^* = \tau_k \ \forall k \in V(j)$$

If $\tau_i > \lambda_i$ for some stockpoints, an additional problem reduction property may hold.

Property 4.3.4. If there exists two stockpoints, r and $a \in A(r)$, that satisfy

$$\sum_{j \in w(a,r)} (\tau_j - \lambda_j) \geq 0 \ and \tag{4.17}$$

$$\sum_{j \in w(a,q)} (\tau_j - \lambda_j) < 0 \ \forall q \in w(a,r) \backslash \{r\}, \tag{4.18}$$

then stockpoint $a \in A$ requires no safety stock coverage and is excluded from further analysis.

Proof: Appendix B, Proposition B.3.2.
If Property 4.3.4 is exploited until no further problem reduction is possible, the second condition is always satisfied. The consideration of a non-reducible convergent system yields the following characterization for $a(i)$.[63]

Property 4.3.5. The maximum predecessor service time results from the path with the largest sum of processing times minus the times implying zero safety stocks.

$$a(i) = \arg \max_{a \in A(i)} \left\{ \sum_{j \in w(a,i)} (\lambda_j - \tau_j) \right\}.$$

---

[63] For a proof see Appendix B, Proposition B.3.3.

As a result of properties 4.3.3-4.3.5, the following property applies.[64]

Property 4.3.6. For a non-reducible convergent problem, coverage times of an optimal policy satisfy

$$T_i \geq \tau_i \qquad\qquad i = 1, ..., n.$$

The solution of a non-reducible convergent network optimization problem under γ-service levels requires the evaluation of the following coverage time values.

$$T_n^* \in \{\lambda_n + \Delta(l) + \sum_{j \in w(l,n)\backslash\{n\}} (\lambda_j - \tau_j), \; l = 1, ..., n\} \qquad \text{and}$$

$$T_i^* \in \left\{ [\lambda_i + \Delta(l) + \sum_{j \in w(l,i)\backslash\{i\}} (\lambda_j - \tau_j) - \sum_{j \in N(i)\backslash\{i\}} (T_j^* - \lambda_j)]^+, \forall l \in V(i) \right\}$$

$$\text{for } i = 1, ..., n-1$$

with $\Delta(1) = 0$ and $\Delta(i) = \max_{v \in N(i)} \left\{ \sum_{j \in w(i,v)} (\tau_j - \lambda_j) \right\}^+ \quad i = 2, ..., n-1.$

The definition of $\Delta(i)$ differs from the one given for serial and divergent systems for notational simplicity. Successive evaluation of coverage times for $i = 1, ..., n-1$ results in the following non-recursive characterization.

$$T_i^* \in \{[\lambda_i + \Delta(l) + \sum_{j \in w(l,i)\backslash\{i\}} (\lambda_j - \tau_j) - [\Delta(r) + \sum_{j \in w(r,n)} (\lambda_j - \tau_j)$$

$$- \sum_{j \in w(n(i),n)} (\lambda_j - \tau_j)]^+]^+, \; \forall l \in V(i)\backslash\{i\}, \; \forall r \in V(n)\}$$

Analogously to α-service levels, the set of optimal coverage time values reduces to the set presented for a serial system that operates under γ-service level constraints. The proposed Dynamic Programming algorithm has to be adjusted only with respect to state and decision spaces which, under γ-service levels and after exploitation of all problem reduction properties, become

$$Z_n = \{0\} \qquad \text{and}$$

$$Z_i = \left\{ \max\{\Delta(n(i)); \; \Delta(n(r)) + \sum_{j \in w(r,n)} (\lambda_j - \tau_j) \right.$$

$$\left. - \sum_{j \in w(n(i),n)} (\lambda_j - \tau_j)\}, \; \forall r \in V(n)\backslash\{n\} \right\} \qquad i = 1, ..., n-1$$

---

[64] For a proof see Appendix B, Proposition B.3.4.

$$U_i(z_i) = \left\{ \max\{\tau_i \; ; \; \lambda_i + \Delta(l) + \sum_{j \in w(l,i) \backslash \{i\}} (\lambda_j - \tau_j) - z_i\}, \; \forall l \in V(i) \right\}$$
$$i = 1, ..., n.$$

4.4 General Systems

In multi-echelon inventory literature, there are only a few approaches directed to safety stock planning for networks of general product structure. For models that operate under the full delay assumption, DE KOK, VISSCHERS[65] generalize the equivalence result obtained for convergent networks to more general systems with component commonality. They show that this special type of network can be transformed into an equivalent network of divergent structure that can be analyzed by the framework summarized in Section 4.2. GRAVES, WILLEMS[66] show how to extend the SIMPSON approach with the underlying no delay assumption to networks that suffice a spanning tree connection between the stockpoints. Nevertheless, there is still no general purpose approach to cope with general network structures. In the following, the SIMPSON model is formulated and analyzed for general systems. After presenting the model formulation and discussing some properties of an optimal solution, exact and heuristic solution procedures are discussed. The sufficiency of the outcome of heuristic procedures is investigated on a test set of 50 different system configurations.

4.4.1 No Delay Optimization Model

In a general multi-echelon system, stockpoints are connected with several predecessors and several successors. Under the assumptions of the SIMPSON model, every stockpoint guarantees an identical service time to its successors. The replenishment lead time of a stockpoint in a general (acyclic) system is affected by the service times of all predecessors in the same way as pointed out for convergent systems.

$$\min \; C = \sum_{i \in A} h_i \cdot \sigma_i \cdot k_i \cdot \sqrt{\lambda_i - ST_i}$$
$$+ \sum_{i \in P} h_i \cdot \sigma_i \cdot k_i \cdot \sqrt{\max_{j \in v(i)} \{ST_j\} + \lambda_i - ST_i}$$
$$+ \sum_{i \in E} h_i \cdot \sigma_i \cdot k_i \cdot \sqrt{\max_{j \in v(i)} \{ST_j\} + \lambda_i}$$

---

[65] DE KOK, A.G., VISSCHERS, J.W.C.H [1999].
[66] GRAVES, S.C., WILLEMS, S.P. [1998].

$$s.t. \qquad 0 \le ST_i \le \lambda_i \qquad\qquad \forall i \in A$$

$$0 \le ST_i \le \max_{j \in v(i)} \{ST_j\} + \lambda_i \qquad \forall i \in P$$

$$ST_i = 0 \qquad\qquad\qquad \forall i \in E.$$

The objective function that minimizes safety stock holding cost is non-linear as already outlined for the three basic types of multi-echelon systems. For the above service time formulation, even the constraints are non-linear due to the max-operator for selecting the longest material availability time. In general, a function defined by the maximum of convex/linear functions is a convex function.[67] In order to gain more insight into properties of an optimal safety stock allocation policy, it is desirable to derive an optimization model formulation with properties similar to the extreme point characterizations found for basic structures. Therefore, an additional problem variable θ_i is introduced for every intermediate and final-stage stockpoint $i \in P \cup E$. This variable represents the maximum time that is required to make all components available under the assumption of no internal delays, that is that every predecessor stockpoint delivers the requested products within its service time.

$$\theta_i = \max_{j \in v(i)} \{ST_j\} \qquad\qquad \forall i \in P \cup E$$

In order to assure that θ_i exactly equals the maximum of all predecessor service times, the additional constraints

$$\theta_i \ge ST_j \qquad\qquad \forall j \in v(i), i \in P \cup E$$

are introduced. Under this transformation, the service time optimization problem for a general acyclic network is given by

$$\min\ C = \sum_{i \in A} h_i \cdot \sigma_i \cdot k_i \cdot \sqrt{\lambda_i - ST_i}$$

$$+ \sum_{i \in P} h_i \cdot \sigma_i \cdot k_i \cdot \sqrt{\theta_i + \lambda_i - ST_i}$$

$$+ \sum_{i \in E} h_i \cdot \sigma_i \cdot k_i \cdot \sqrt{\theta_i + \lambda_i}$$

$$s.t. \qquad 0 \le ST_i \le \lambda_i \qquad\qquad \forall i \in A$$

$$0 \le ST_i \le \theta_i + \lambda_i \qquad\quad \forall i \in P$$

$$ST_i = 0 \qquad\qquad\qquad \forall i \in E$$

$$\theta_i \ge ST_j \qquad\qquad \forall j \in v(i), i \in P \cup E.$$

For an optimal policy, $\theta_i = ST_j$ will hold for (at least) one $j \in v(i)$. Otherwise, the availability time variable θ_i can be reduced and this reduction implies a lower objective function value. The extended formulation with additional

---

[67] See SYDSAETER, K., HAMMOND, P.J. [1995].

variables and constraints represents a concave minimization problem under linear constraints and, therefore, resembles the same extreme point properties exploited for serial, divergent, and convergent networks. In order to express the optimization model in terms of coverage time variables, the substitution of

$$ST_i = \max_{j \in v(i)} \{ST_j\} + \lambda_i - T_i$$

yields an equivalent coverage time optimization model.

$$\min \ C = \sum_{i=1}^{n} h_i \cdot \sigma_i \cdot k_i \cdot \sqrt{T_i}$$

$$s.t. \qquad \max_{w \in W(A,i)} \left\{ \sum_{j \in w} (\lambda_j - T_j) \right\} \geq 0 \qquad i = 1, ..., n$$

$$\sum_{j \in w} T_j \geq \sum_{j \in w} \lambda_j \qquad \forall w \in W(A, E)$$

$$T_i \geq 0 \qquad i = 1, ..., n.$$

The additional complexity that arises for general multi-echelon systems in comparison to the basic network types concerns the numerical characterization of possible coverage time values for an optimal allocation policy. The set of coverage time candidates for divergent and serial systems can be characterized by a forward logic, that is that the set of coverage time candidates can be described only depending on upstream coverage times, and for convergent and serial systems by a backward logic, that is that the set of coverage time candidates can be described only depending on downstream coverage times. For general network structures, this result does not hold and the set of coverage time candidates depends on both, upstream and downstream coverage times.

The constraints that define the set of feasible coverage time allocations are influenced by three types of constraints:

- a non-negativity constraint for each coverage time,
- cumulative coverage of processing times for every path through the network,
- non-negativity of replenishment lead times.

The joint analysis of these three types provides the following characterization for a coverage time.

$$T_i^* = \left(\lambda_i + \underbrace{\max_{w \in W(A,i)} \left\{ \sum_{j \in w \setminus \{i\}} (\lambda_j - T_j^*) \right\}}_{L(i)} - \underbrace{\min_{w \in W(i,E)} \left\{ \sum_{j \in w \setminus \{i\}} (T_j^* - \lambda_j) \right\}}_{M(i)} \right)^{+}$$

This characterization of T_i^* is influenced by upstream and downstream decisions on coverage times. Upstream decisions determine the replenishment lead time $L(i)$. The requirement that on every downstream path to any final product the cumulative processing time has to be covered can imply an excess coverage to i defined as the minimal excess coverage $M(i)$ of all downstream paths. Finally, the max-argument ensures the non-negativity of coverage time values. The interpretation of relevant coverage time values is similar to the ones given for the basic network structures, though it is not possible to characterize the values purely by upstream or downstream decisions. A "cover all or nothing" policy implies that, in general, a stockpoint will cover zero time or the entire replenishment lead time $L(i)$. Nevertheless, the simultaneous impact of downstream coverage decisions which might provide a minimal excess coverage for all paths from i to the final stage has to be taken into account by reducing coverage requirements that result from the replenishment lead time by the minimal excess $M(i)$. Therefore, the relevant decisions at a stockpoint, given the upstream decisions, are zero or equal to the lead time. The relevant decision values, given the downstream coverage times, are to cover cumulative processing times of a path from i to all predecessors of i and subtract the excess that is provided by downstream decisions.

4.4.2 Exact Solution Algorithms

The concave minimization problem can be solved by general purpose concave minimization algorithms.[68] These methods can be classified into

- Cutting plane algorithms,
- Successive approximation methods,
- Branch & Bound techniques.

Other methods that exploit the extreme point based combinatorial "all or nothing" aspect of the problem are

- Complete enumeration,
- Dynamic Programming.

As outlined in the previous section, a straightforward extension of Dynamic Programming algorithms that are developed for divergent and convergent systems is not possible. The simultaneous impact of upstream and downstream decisions and the impossibility of a one sided characterization of candidate coverage times makes the application of Dynamic Programming techniques with single variable states and decisions impossible. The most complex type of general multi-echelon system, to which the single-state variable Dynamic Programming logic applies, is a spanning tree network of stockpoints. In a spanning tree network, two stockpoints are connected by, at most, one path

---

[68] A detailed overview and description of these methods is presented in HORST, R., TUY, H. [1996].

of arcs. For this type of network, GRAVES, WILLEMS[69] present an algorithm for finding an evaluation sequence of the functional equations that ensures a single state variable representation.

In extension to the algorithms presented for divergent and convergent systems, both a forward recursion and a backward recursion Dynamic Programming algorithm can be applied to solve the safety stock coverage time problem for general networks that are more complex than spanning trees. The additional complexity of the problem leads to higher dimensional state and decision spaces. In order to perform the algorithm, a sequence in which the different stockpoints are evaluated must be determined. Therefore, stockpoints are assigned to decision stages. To get a feasible assignment, stockpoints that belong to the same stage are not allowed to be connected, that is stockpoints that are supplied by a stockpoint i at stage l are assigned to stages $k > l$. In general, there are several different assignments of stockpoints to stages that fulfill the above condition and the assignment itself represents an optimization problem with respect to the resulting computational complexity because the assignment determines the sizes of state and decision vectors at each stage. When the number of stages is maximized, that is the minimum number of stockpoints is assigned to a stage, this results in small decision but large state vectors. Minimizing the number of stages, that is that the maximum number of stockpoints is assigned to a stage, results in small state but large decision vectors.

If the connection between stockpoints is characterized by a spanning tree, stockpoints can be evaluated sequentially with the consequence that the state can be characterized by a single variable. With increasing interconnection of stockpoints, additional state variables must be introduced. Nevertheless, for most applications, the number of additional state variables will be considerably small and therefore, the alternative with as many stages as possible seems to be preferable. Note that the successive evaluation of stockpoints in divergent and convergent systems is possible even if they have the same successor or predecessor. For the second assignment alternative in a pure divergent or convergent system, this implies a multi dimensional decision evaluation, though the evaluation is possible successively with moderate computational complexity.

Let $l = 1, ..., \Lambda$ denote the introduced decision stages. Stockpoints assigned to stage l are contained in the set $H(l)$ with $m(l)$ elements. The decision vector $\underline{u}_l = (u_{1,l}, ..., u_{m(l),l})$ represents all coverage decisions for stage l which must be made simultaneously. The definition of states, state transition equations, state and decision spaces, as well as the definition of functional values and functional equations, depends on the direction of the Dynamic Programming algorithm.

---

[69] GRAVES, S.C., WILLEMS, S.P. [1998].

Forward Recursion. The cost function is directed to upstream decision stages and the state incorporates decisions of downstream stages. The state variables represent excess coverage guaranteed by stockpoints being assigned to downstream stages. Since downstream decisions also influence excess coverage of upstream stockpoints, a state variable has to be introduced for all upstream stockpoints at stages $k < l$ with direct downstream successors on stages $k > l$.

- $\underline{z}_{l,l} = (z_{l,l,i})_{i=1,\ldots,m(l)}$: minimal excess coverage for i^{th} stockpoint of stage l

- $\underline{z}_{l,i} = (z_{l,i,j})_{j=1,\ldots,m(i)}$: minimal excess coverage on stages $l + 1, \ldots, \Lambda$ concerning stockpoints assigned to stage i ($i = 1, \ldots, l - 1$), to be introduced for all stockpoints in $H(1) \cup \ldots \cup H(l-1)$ with direct successors in $H(l+1) \cup \ldots \cup H(\Lambda)$

The complete state vector for stage l then is given by $\underline{z}_l = (\underline{z}_{l,l}, \underline{z}_{l,l-1}, \ldots, \underline{z}_{l,1})$. For a given state $\underline{z}_l$ and a decision vector $\underline{u}_l$, the resulting state vector for the preceding stage $l - 1$ in the forward recursion is given by

$$z_{l-1,i,k} = \min_{j \in n(k) \cap H(l)} \{z_{l,i,k}; z_{l,l,j} + u_j - \lambda_j\}.$$

This formulation contains three cases for the state of stockpoint $k \in H(i)$. If all successors of k are assigned to stages $l + 1, \ldots, \Lambda$, minimal excess coverage is not affected by the decision at stage l and $z_{l-1,i,k} = z_{l,i,k}$ holds. If k has a single successor j assigned to l, as it is the case in a serial or convergent system, excess coverage is defined by excess coverage to j plus the coverage decision of j minus the corresponding processing time. The third case, introduced by the non-spanning tree property of a general network, requires one to evaluate minimal excess coverage over all successors on stage l together with the excess resulting from downstream decisions.

- State space
  The state space for the final decision stage is $Z_\Lambda = \{\underline{z}_{\Lambda,\Lambda} = \underline{0}\}$. The state spaces of all other stages are rather complex. One possibility is to generate the respective spaces by recursive generation from the following stage state space and the respective decision spaces.
- Decision space
  The decision space in general is spanned by the individual decision variable spaces that themselves follow the logic of covering cumulative processing times to all predecessor stockpoints.

The value function $f_l(\underline{z}_l)$ represents the optimal costs for stages $1, \ldots, l$ if stage l faces state $\underline{z}_l$. The optimal costs result from the optimal decision at stage l with respective direct inventory cost and optimal costs of upstream stages. Therefore, the functional equation is given by

$$f_l(\underline{z}_l) = \min_{\underline{u}_l \in U_l(\underline{z}_l)} \{c(\underline{u}_l) + f_{l-1}(\underline{z}_{l-1})\} \qquad \forall \underline{z}_l \in Z_l.$$

For the first stage in the forward recursion process, the functional values are

$$f_1(\underline{z}_1) = \sum_{k \in H(1)} c_k((\lambda_k - z_{1,k})^+) \qquad \forall \underline{z}_1 \in Z_1.$$

The safety stock holding costs that are induced by the coverage decision vector $\underline{u}_l$ at stage l are given by

$$c(\underline{u}_l) = \sum_{j \in H(l)} c_j(u_j).$$

Backward Recursion. For the backward algorithm, the cost function is directed to downstream decision stages and the state incorporates decisions of upstream decision stages. State variables represent material availability lead times, i.e. maximum predecessor service times. In order to obtain the respective replenishment lead times, processing times must also be added. Since upstream coverage decisions influence lead times of downstream points, a state variable has to be introduced for all downstream stockpoints at stages $> l$ with upstream predecessors on stages $< l$.

- $\underline{z}_{l,l} = (z_{l,l,k})_{k=1,\ldots,m(l)}$: material lead time for k^{th} stockpoint of stage l
- $\underline{z}_{l,i} = (z_{l,i,k})_{k=1,\ldots,m(i)}$: material lead time of stockpoints on stages $1, \ldots, l - 1$ that supply stockpoints assigned to stage i $(i = l + 1, \ldots, \Lambda)$, to be introduced for all stockpoints in $H(l + 1) \cup \ldots \cup H(\Lambda)$ with direct predecessors in $H(1) \cup \ldots \cup H(l - 1)$

The complete state vector for stage l is given by $\underline{z}_l = (\underline{z}_{l,l}, \underline{z}_{l,l+1}, \ldots, \underline{z}_{l,\Lambda})$. For a given state $\underline{z}_l$ and decision vector $\underline{u}_l$ for stage l, the resulting state vector for the succeeding decision stage $l + 1$ in the backward recursion is

$$z_{l+1,i,k} = \max_{j \in v(k) \cap H(l)} \{z_{l,i,k}; z_{l,l,j} - u_j + \lambda_j\}.$$

Corresponding with the forward recursion, this formulation contains three cases for the state with respect to stockpoint $k \in H(i)$. If all predecessors of k are assigned to stages $1, \ldots, l-1$, the material lead time is not affected by the decision at stage l and $z_{l+1,i,k} = z_{l,i,k}$. If k has a single predecessor j assigned to stage l, as it is the case for a serial or divergent system, the material lead time is defined by the material lead time to j plus the processing time of j minus the corresponding coverage time. The third case, introduced by the non-spanning tree property of a general network, requires one to evaluate the material lead time over all predecessors on stage l together with the material lead time resulting from upstream decisions.

- State space
  For the first stage, the state space is $Z_1 = \{(z_{1,k})_{k=1,\ldots,m(1)} = \underline{0}\}$. For succeeding stages, the respective spaces are generated by the cover all or nothing property of upstream decisions.

- Decision space
  The individual decision spaces that span the overall space mainly result from the states under the additional "cover all or nothing" property.

The value function $f_l(z_l)$ represents the optimal costs for stages $l+1, ..., \Lambda$ if stage l faces state z_l. The optimal costs result from optimal decision at stage l with the respective direct cost and optimal costs of downstream stages under the state transition.

$$f_l(z_l) = \min_{u_l \in U_l(z_l)} \{c(u_l) + f_{l+1}(z_{l+1})\} \qquad \forall z_l \in Z_l$$

For the final stage in the forward recursion, the functional values are

$$f_\Lambda(z_\Lambda) = \sum_{k \in H(\Lambda)} c_k(z_{\Lambda,k}) \qquad \forall z_\Lambda \in Z_\Lambda.$$

The choice between the implementation of a forward or backward recursion method depends on the dominant type of stockpoint interrelation. If the supply chain network is dominated by divergent problem aspects, a backward recursion is preferable, whereas, a convergent dominance makes a forward algorithm more advantageous. Such an assignment provides a minimum number of state variables and often the possibility of an independent minimization with respect to the decision variables at a stage. Note that additional state variables are necessary if a convergent system is optimized by a backward or if a divergent system is optimized by a forward recursion method.

4.4.3 Heuristic Solution Algorithms

Heuristic methods for solving the safety stock allocation problem can be classified into simultaneous solution methods without an optimality guarantee, for example a solution of the problem with a linearized objective function, solution construction methods to obtain a feasible solution, and solution improvement techniques.

4.4.3.1 Linear Approximation. The only non-linear component of the optimization problem is given by the square root of safety stock coverage times in the objective function. One technique for a heuristic solution of separable non-linear optimization problems is the approximation by a Linear Program which can be solved efficiently by using standard software packages. The smaller the degree of non-linearity/concavity of the objective function, the smaller will the error of this approximation be. The computational complexity on the one hand and the solution quality on the other hand highly depend on the number r_i of reference points introduced for the corresponding coverage time. The interval of relevant values for the coverage times T_i is

$$0 \le T_i \le T_i^{max}, \qquad T_i^{max} = \max_{w \in W(A,i)} \sum_{j \in w} \lambda_j.$$

The most simple case is the one with two reference points for each stockpoint, i.e. $r_i^1 = 0$, $r_i^2 = T_i^{max}$. Then, the linear slope of the objective function with respect to T_i is given by

$$c_i = \frac{h_i \cdot \sigma_i \cdot k_i}{\sqrt{T_i^{max}}}$$

for the safety stock holding cost criterion. For the on-hand stock criterion the approach is analogous, only the corresponding on-hand stocks have to be determined. Linear approximations underestimate the true objective function value as depicted in Figure 4.3.

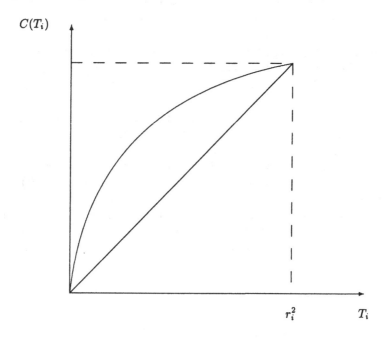

Fig. 4.3. Linear approximation of the objective function

The approximate Linear Program is given by

$$\min\ C = \sum_{i \in A} c_i \cdot (\lambda_i - ST_i) + \sum_{i \in P} c_i \cdot (\theta_i + \lambda_i - ST_i) + \sum_{i \in E} c_i \cdot (\theta_i + \lambda_i)$$

$$
\begin{array}{lll}
s.t. & 0 \leq ST_i \leq \lambda_i & \forall i \in A \\
& 0 \leq ST_i \leq \theta_i + \lambda_i & \forall i \in P \\
& ST_i = 0 & \forall i \in E \\
& \theta_i \geq ST_j & \forall j \in v(i), i \in P \cup E.
\end{array}
$$

The incorporation of more than two reference points requires additional variables and constraints.[70]

4.4.3.2 Solution Characterization. Solution improvement techniques like (Steepest) Descent, Simulated Annealing, Threshold Accepting, and Tabu Search to be presented in the following share a common feature. First, a starting solution has to be obtained by some solution construction algorithm. Small perturbations yield different solutions close to the actual solution. The cost quality of the perturbed solution is compared to the actual solution and, based on the comparison result, it is decided whether the new solution is accepted as the basis for further perturbations. In order to perform this iterative solution inspection process, a solution representation, the perturbation mechanism, and a solution evaluation method have to be defined.

Solution Representation. The solution to the safety stock allocation problem allows for two obvious characterizations. One representation simply uses the vector $\underline{T}$ of coverage times. On the other hand, regarding the extreme point property, a solution is also characterized by the set of stockpoints that hold safety stocks and by the set of those that operate as inventory-less installations. This representation strongly relates to the combinatorial characteristic observed for optimal policy candidates. By introducing a binary representation $\underline{b}$, a single bit $b(i) \in \{0, 1\}$ indicates if stockpoint i will hold safety stocks ($b(i) = 1$), or if the safety stock will be zero ($b(i) = 0$). Since periodic review of the inventory status always requires for a positive safety stock coverage for all final-stage stockpoints $i \in E$, the binary representation only includes a bit for stockpoints $i \in A \cup P$. Therefore, the solution representation used for the following considerations is given by $\underline{b} = (b(i))_{i=1,...,n-|E|}$ with

$$b(i) = \begin{cases} 1 \text{ if safety stocks are held at } i \\ 0 \text{ otherwise} \end{cases}.$$

In total, the binary representation yields $2^{n-|E|}$ alternative solutions. Nevertheless, in terms of coverage time requirements that result from the binary safety stock indicator, some of these policies that are different in their representation will result in identical coverage time requirements.

Definition of Perturbations. With the use of a binary solution representation, a customary definition of the neighborhood to a given solution (defined as the set of all perturbations being possible to a specific solution) is to include all zero/one switches for every single bit within the binary vector.

$$N(\underline{b}) = \{\hat{\underline{b}} | b(i) = 1 - \hat{b}(i) \text{ for one } i \in A \cup P, b(j) = \hat{b}(j) \ \forall j \in A \cup P \backslash \{i\}\}.$$

Then, the size of the neighborhood to a given solution is $n - |E|$.

---

[70] HORST, R., TUY, H. [1996].

Solution Evaluation. In order to evaluate a given indicator vector, the corresponding safety stock coverage time vector must be determined. With the coverage time and the equivalent safety stock, the value of the cost function can be computed. Recalling the extreme point property, each candidate for an optimal coverage time can be characterized by

$$T_i = \Big(L(i) - M(i)\Big)^+ \qquad i = 1, ..., n.$$

For a given indicator vector $\underline{b}$, this expression is unique and can be evaluated. From the interpretation of the set of relevant coverage and service times, it is obvious that $L(i)$ can be evaluated independent of $M(i)$. Therefore, the evaluation is decomposed into two steps. In a first step, replenishment lead times for the solution representation are determined and in a second step, the resulting excess coverage time spans and the corresponding coverage time values are derived. The component $L(i)$ can be determined starting with the first-stage stockpoints. There, the replenishment lead time is identical to the processing time λ_i. Proceeding with a stockpoint where all predecessors have already been evaluated, the replenishment lead time consists of the processing time belonging to this stockpoint and the maximum uncovered time of all direct predecessors. If a predecessor j holds safety stocks ($b(j) = 1$), then the service time of j is zero and has no impact on the replenishment lead time of i. Therefore, $L(j)$ is multiplied by zero $(1 - b(j))$. In the opposite case ($b(j) = 0$), $L(j)$ may influence the replenishment lead time of i if it determines the maximum value. Therefore, $L(j)$ is multiplied by one $(1 - b(j))$.

$$L(i) = \qquad \lambda_i \qquad\qquad \forall i \in A$$
$$L(i) = \lambda_i + \max_{j \in v(i)} \{L(j) \cdot (1 - b(j))\} \ \forall i \in P \cup E$$

Note that this evaluation can be performed without knowing the exact coverage time values in advance. The excess coverage of downstream stockpoints is computed starting with the stockpoints without successors. By definition, the excess for these stockpoints is zero. The other stockpoints are evaluated in a sequence where the next stockpoint to be evaluated faces only successors with already determined coverage time value. The determination of the excess value depends on coverage time values of all direct successors.

$$M(i) = \qquad 0 \qquad\qquad \forall i \in E$$
$$M(i) = \min_{j \in n(i)} \{T_j - \lambda_j + M(j)\} \ \forall i \in A \cup P$$

When replenishment lead time $L(i)$ and excess coverage $M(i)$ are known, the resulting coverage time requirement T_i is computed, which is a necessary input to determine excess expressions for all predecessors of i. After all coverage times $\underline{T}(\underline{b})$ have been derived from the binary safety stock configuration vector, the associated costs are given by

$$C(\underline{b}) = C(\underline{T}(\underline{b})) = \sum_{i=1}^{n} h_i \cdot \sigma_i \cdot k_i \cdot \sqrt{T_i(\underline{b})}.$$

4.4.3.3 Solution Construction Techniques. Solution construction methods for general systems follow the same lines introduced for the generation of feasible initial solutions in a serial system.

- Random solution generation
- Rules of thumb
  - (a) end-item buffering
  - (b) first- and final-stage buffering
  - (c) all item buffering
- Successive assignment

Solutions obtained from these construction heuristics can be quite unsatisfactory. Random solution generation does not take into account any cost information and, therefore, only makes sense together with a solution improvement algorithm. In principle, the same argument holds for rules of thumb. Nevertheless, value added based inventory holding cost with ongoing processing steps and the square root law give reasons that rule (a) and (b) can generate reasonable approximations. Successive assignment of coverage times to stockpoints takes into account the direct cost impact of the assignment but this principle neglects the impact of a coverage decision on upstream coverage times. Therefore, it is reasonable to apply solution improvement techniques to initial solutions obtained by the above methods.

4.4.3.4 Descent Solution Improvement. The basic solution improvement algorithms are descent methods. The idea behind this approach is to accept only solution improving perturbations. The algorithm iterates further solution perturbations until all neighboring solutions to the actual one are worse. Two algorithmic variants to this idea are the descent and the steepest descent method.

Descent Method. The descent method evaluates the cost value $C(\underline{b}^j)$ of one neighbor solution $\underline{b}^j$ and accepts this solution if the cost difference ΔC to the value $C(\underline{b}^i)$ of the actual solution is negative. In the case $\Delta C > 0$ (the neighbor provides no solution improvement) the next neighbor to the actual solution is evaluated until an improving neighbor is found or the stopping criterion, that all neighbors to the actual solution are worse, is fulfilled.

> **procedure** descent
>   **begin**
>   select an initial configuration $\underline{b}^i$;
>   **repeat**
>   generate a neighbor $\underline{b}^j$ of configuration $\underline{b}^i$;
>   calculate $\Delta C := C(\underline{b}^i) - C(\underline{b}^j)$;
>   **if** $\Delta C > 0$ **then** $\underline{b}^i := \underline{b}^j$;
>   **until** $C(\underline{b}^j) \geq C(\underline{b}^i)$ for all neighbors $\underline{b}^j \in NB(\underline{b}^i)$;
>   **end**;

Besides the determination of a starting solution, there is one degree of freedom concerning the choice of the next neighbor to be evaluated. The candidate can be chosen in either a fixed or a random order. For the choice in a fixed order, neighbors that are generated from a switch of a safety stock indicator bit can be evaluated with (a) increasing or (b) decreasing stockpoint index. The random choice of a neighbor requires an additional list. This list stores the already evaluated neighbors since the last change of the actual solution took place in order to avoid double inspections and also to be able to check the stopping criterion that all neighbors to the actual solution have been evaluated before terminating the iteration.

Steepest Descent. The difference between the steepest descent variant and the simple descent idea concerns the number of evaluated neighbors before deciding upon the change of the actual solution. Where the descent method evaluates one neighbor at a time, the steepest descent method evaluates all $n - |E|$ neighbors to the actual solution and returns the best neighbor. If this result yields a solution improvement, the best neighbor becomes the actual solution. Otherwise, the iteration terminates because there are no improving neighbors to the actual solution.

> **procedure** steepest descent
> **begin**
> select an initial configuration $\underline{b}^i$;
> **repeat**
> calculate $\Delta C := \min\limits_{\underline{b}^j \in NB(\underline{b}^i)} \{C(\underline{b}^j)\}$;
> **if** $\Delta C < C(\underline{b}^i)$ **then** $\underline{b}^i := \arg\min_{\underline{b}^j \in NB(\underline{b}^i)} \{C(\underline{b}^j)\}$;
> **until** $C(\underline{b}^j) > C(\underline{b}^i)$ for all neighbors $\underline{b}^j \in NB(\underline{b}^i)$;
> **end;**

The disadvantage of local improvement procedures is that they might get trapped in local minima. The stopping criterion excludes the possibility that local minima can be left. A simple possibility to overcome this shortcoming is to start the algorithm from a set of different initial solutions and to use the best solution found within these multiple local search runs. The small computation time requirements of each descent algorithm together with the law of the big number give rise to the conjecture that this approach might provide reasonable solutions. On the other hand, this search is neither guided nor satisfactory from a theoretical point of view. More theory based approaches towards the local minima problem are presented in the next three sections. Simulated Annealing accepts an inferior solution with some probability in order to leave local minima, whereas the deterministic variant Threshold Accepting accepts every solution with a disadvantage not larger than a threshold value. A different idea is followed by Tabu Search. This approach works with a dynamic restricted neighborhood where solutions visited within a given span of iterations are forbidden to be revisited.

4.4.3.5 Simulated Annealing. Simulated Annealing uses the descent method as a main building block. Beginning with an initial solution, neighbor solutions are generated and evaluated with respect to their cost difference to the actual solution. In accordance with the descent method, an improving solution is accepted and used as the actual solution for following evaluations. In order to escape from local minima and also to explore wider regions of the solution space, inferior solutions are accepted by a probability mechanism. The probability that a solution is accepted increases with decreasing cost difference ΔC to the actual solution. With ongoing iterations, this probability mechanism is modified and the solution acceptance probability decreases. From the analogy of the Simulated Annealing algorithm to physics, this modification is often called cooling of the accepting temperature TP_c. The effect is that even large disimprovements are accepted in the first iterations, whereas the algorithm converges to the descent method in later iteration steps.

$$P\{\text{a disimproving solution is accepted}\} = e^{-\Delta C/TP_c}$$

Therefore, small (positive) cost deviations ΔC as well as a high temperature TP_c provide higher acceptance probabilities and vice versa. In order to enable a better comparison among all local search techniques, the Simulated Annealing algorithm implemented is a variant that stores the best solution found throughout all evaluations. With these prerequisites, the following basic algorithm[71] is suggested to find solutions to the safety stock coverage time allocation problem.

procedure simulated annealing
    begin
        select an initial configuration $\underline{b}^i$;
        $\underline{b}^* := \underline{b}^i$; $C^* := C(\underline{b}^i)$;
        select an initial temperature TP_0;
        set iteration counter $c := 0$;
        repeat
            for $n := 1$ **to** PL **do begin**
                generate a neighbor $\underline{b}^j$ of configuration $\underline{b}^i$;
                calculate $\Delta C := C(\underline{b}^j) - C(\underline{b}^i)$;
                if $\Delta C < 0$
                then $\underline{b}^i := \underline{b}^j$; **if** $C(\underline{b}^j) < C^*$ **then** $\underline{b}^* := \underline{b}^j$; $C^* := C(\underline{b}^j)$;
                else if $e^{-\Delta C/TP_k} > random$ **then** $\underline{b}^i := \underline{b}^j$;
            end;
            $TP_{c+1} := a \cdot TP_c$;
            $c := c + 1$;
        until stopping criterion is fulfilled;
    end;

---

[71] See EGLESE, R.W. [1990] for a tutorial on Simulated Annealing.

Besides the mechanism for generating neighbors as already mentioned for the descent algorithm, four generic parameters are included in the formulation and appropriate values must be determined. The choice has a strong influence on the performance of the algorithm and represents an optimization problem itself. Therefore, different parameter settings are tested in Section 4.5.3 in order to give guidelines for appropriate values. In the following, assume that the initial configuration is given by the end-item buffering strategy.

(1) Initial temperature TP_0

The initial temperature indicates to which extent inferior solutions are accepted at the beginning of the annealing algorithm. In order to allow for a wide range search, this temperature should not be too low since the local minimum problem might appear again. On the other hand, if the temperature is too high, almost every solution is accepted and the algorithm converges to a random search.

(a) A first possibility to initialize the temperature is to choose its value in a way that a solution with the double objective function value to the end-item buffering opportunity $C(\underline{b}^{EEB})$ is accepted with probability P_1.

$$e^{-\frac{C(\underline{b}^{EEB})}{TP_0}} = P_1 \Leftrightarrow TP_0 = -\frac{C(\underline{b}^{EEB})}{\ln P_1}$$

A reasonable interval of values for P_1 is given by $P_1 \in (0, ..., 1)$.

(b) An alternative choice for the initial temperature is to determine the best and the worst solution within the neighborhood of the initial configuration in order to get an impression on the range of cost values and to set the initial temperature to some multiple P_1 of the difference of both cost values.

$$TP_0 = P_1 \cdot \left(\max_{\underline{b}^j \in NB(\underline{b}^{EEB})} \{C(\underline{b}^j)\} - \min_{\underline{b}^j \in NB(\underline{b}^{EEB})} \{C(\underline{b}^j)\} \right)$$

(2) Temperature function

The temperature function (cooling schedule) describes the extent to which inferior solutions are accepted in each iteration. Here, the quite common proportional cooling schedule with cooling factor P_2 is used.

$$TP_{c+1} = P_2 \cdot TP_c$$

If the temperature is lowered too fast, the algorithm tends to get trapped in unfavorable regions of the solution set.

(3) Number PL of evaluated neighbors at each temperature

A sufficiently large number of solutions has to be evaluated before lowering the temperature. It is assumed that the size of PL is a multiple P_3 of the neighborhood size $n - |E|$. If PL is too small, the solution space might not be sufficiently explored, whereas if it is too large, computation time increases without guiding the search into new promising regions.

$$PL = P_3 \cdot (n - |E|)$$

(4) Stopping criterion

There are different stopping criteria either based on the number of iterations or on the performance of the improvement process.

(a) In general, with the temperature approaching zero, the algorithm converges to the descent method. Since the temperature never reaches zero under a proportional cooling schedule, a small temperature value has to be determined where the algorithm terminates. When objective function values are not normalized to a zero/one interval, the behavior of cost values and the temperature value are interrelated. In order to introduce some normalization, a ratio between the best solution value found $C(\underline{b}^*)$ and a benchmark solution $C(\underline{b}^{EEB})$ can be utilized.

$$TP_k \le P_4 \cdot C(\underline{b}^*)/C(\underline{b}_{EEB}).$$

(b) Terminate the algorithm, if there was no improvement of the best solution $\underline{b}^*$ within the last P_4 iterations.

(c) Terminate the algorithm after a fixed number of cooling steps $P_4 \cdot (n - |E|)$.

4.4.3.6 Threshold Accepting. The probabilistic element of accepting disimproving solutions enables the Simulated Annealing algorithm to escape from local cost minima. A deterministic variant of this idea characterizes the Threshold Accepting algorithm.[72] Instead of accepting disimproving solutions with some probability, every inferior solution with a cost difference smaller than a given threshold value TH is accepted. The function of the threshold value is lowered in a way similar to the one discussed for the temperature function.

The choice of the generic parameters is similar to Simulated Annealing. Therefore, the same discussion applies here. With respect to solution acceptance, Dueck[73] discusses two additional variants to the Threshold Accepting algorithm. The criterion used by Threshold Accepting is based on the relative difference $\Delta C = C(\underline{b}^j) - C(\underline{b}^i)$ between the actual solution and its neighbor. The variant Great Deluge substitutes this relative criterion by an absolute criterion. Every solution that is below a cost benchmark $\overline{C}$ is accepted. With increasing iterations this cost benchmark is lowered. The second variant Record-to-Record-Travel uses a relative deviation criterion. However, instead of comparing the neighbor quality to the actual solution, it is compared to the best solution (record) found in the previous iterations of the algorithm.

---

[72] DUECK, G., SCHEUER, T. [1990].
[73] DUECK, G. [1993].

procedure threshold accepting
   begin
        select an initial configuration $\underline{b}^i$;
        $\underline{b}^* := \underline{b}^i$; $C^* := C(\underline{b}^i)$;
        select an initial threshold TH_0;
        set iteration counter $c := 0$;
        repeat
          for $n := 1$ **to** PL **do begin**
            generate a neighbor $\underline{b}^j$ of configuration $\underline{b}^i$;
            calculate $\Delta C := C(\underline{b}^j) - C(\underline{b}^i)$;
              if $\Delta C < TH_c$
                then $\underline{b}^i := \underline{b}^j$; **if** $C(\underline{b}^j) < C^*$ **then** $\underline{b}^* := \underline{b}^j$; $C^* := C(\underline{b}^j)$;
          end;
          $TH_{c+1} := a \cdot TH_c$;
          $c := c + 1$;
        until stopping criterion is fulfilled;
  end;

4.4.3.7 Tabu Search. A different approach to the local minimum problem is pursued by Tabu Search.[74] Revisiting a solution is avoided by forbidding already evaluated neighbors. Therefore, Tabu Search works with a neighborhood definition that is varying from iteration to iteration. The second main feature is taken from the steepest descent algorithm. The criterion for selecting the next solution is to choose the best non forbidden neighboring solution. In contrast to steepest descent, the best neighbor can even be the best disimproving solution in order to leave a local minimum.

For restricting the neighborhood definition to unevaluated solutions, it is necessary to store all solutions visited in previous iterations. In every iteration it has to be checked, which solutions of the original neighborhood are forbidden. This straightforward implementation of the Tabu Search idea can require a large amount of storage space and computation time. Therefore, the general idea is simplified. The transition from one solution to a neighbor is called a move. Under the safety stock allocation definition of a solution characterization, a move is characterized by a single attribute, for instance by the stockpoint for which the safety stock coverage decision is converted. Instead of storing all previous solutions, only the attributes of applied moves are memorized. Due to the fact that a reverse move (applied at least two iterations later) must not necessarily result in an already evaluated solution, the forbidden moves are stored only for a given number of iterations. Therefore, in the basic variant, the tabu list of forbidden moves has a constant length.

---

[74] For a tutorial see GLOVER, F. [1989], [1990], [1993].

procedure tabu search
    begin
        select initial configuration $\underline{b}^i$;
        $\underline{b}^* := \underline{b}^i$; $C^* := C(\underline{b}^i)$;
        initialize tabu list TL;
        set iteration counter $c := 0$;
        repeat
          $\underline{b}^i := \arg\min_{\underline{b}^j \in NB(\underline{b}^i)\backslash TL}\{C(\underline{b}^j)\}$;
          if $C(\underline{b}^i) < C^*$ **then** $\underline{b}^* := \underline{b}^i$; $C^* := C(\underline{b}^i)$;
          $c := c + 1$;
          update tabu list TL;
        until stopping criterion is fulfilled;
    end;

An aspiration level criterion addresses forbidden but favorable solutions. Because moves instead of solutions are forbidden, favorable solutions within the neighborhood of the actual solution cannot be chosen. To overcome this drawback, an additional aspiration level criterion can be introduced. If a forbidden solution satisfies this criterion, it is accepted in spite of being forbidden. A very popular aspiration level criterion is that the forbidden solution is better than a threshold, often set equal to the best solution found up to the present iteration[75].

procedure tabu search with aspiration criterion
    begin
        select initial configuration $\underline{b}^i$;
        $\underline{b}^* := \underline{b}^i$; $C^* := C(\underline{b}^i)$;
        initialize tabu list TL;
        set iteration counter $c := 0$;
        repeat
          $\underline{b}^l := \arg\min_{\underline{b}^j \in NB(\underline{b}^i)\backslash TL}\{C(\underline{b}^j)\}$;
          $\underline{b}^m := \arg\min_{\underline{b}^j \in TL}\{C(\underline{b}^j)\}$;
          if $C(\underline{b}^m) < C(\underline{b}^l)$ **and** $C(\underline{b}^m) < C^*$
              then $\underline{b}^i := \underline{b}^m$; $\underline{b}^* := \underline{b}^m$; $C^* := C(\underline{b}^m)$
              else $\underline{b}^i := \underline{b}^l$; **if** $C(\underline{b}_l) < C^*$ **then** $\underline{b}^* := \underline{b}^l$; $C^* := C(\underline{b}^l)$;
          $c := c + 1$;
          update tabu list TL;
        until stopping criterion is fulfilled;
    end;

Each of the two Tabu Search procedures contains two generic parameters, the size of the list of forbidden moves and the stopping criterion.

(1) Stopping criterion
    Most stopping criteria are either based on the number of iterations or on some measure of solution quality improvement.

---

[75] See record-to-record travel.

(a) Number of iterations
The algorithm terminates if the number of iterations exceeds a multiple P_1 of the neighborhood size $n - |E|$.
(b) Solution quality improvement
Terminate the iteration process if no new best solution was found within the last P_1 iterations.
(2) Tabu list
In the presented Tabu Search framework with a static, FIFO updated tabu list, the only parameter is the size which is set to

$$|TL| = P_2 \cdot (n - |E|).$$

4.5 Comparative Analysis of Heuristic Solution Procedures

In this section, the heuristic procedures presented and developed in Section 4.4.3 are parameterized and compared. For most of the modern local search solution improvement techniques, it is necessary to find reasonable parameters in order to achieve a satisfactory solution quality.

4.5.1 Test Set Generation

For the purpose of testing the numerical solution algorithms, a set of different system configurations is randomly generated. The reason for generating the test data randomly is justified by the fact that the algorithms are developed to assist the stock norm determination of practical applications. For large and complex systems, a factorial test set design where all possible combinations of parameters are analyzed is prohibitive. On the other hand, to analyze the performance based on some typical application data sets is rather limited.
Network Structure
In the following, the connections between the stockpoints are generated. Without loss of generality, it is assumed that material requirements coefficients $a_{i,j}$ are equal to one if two stockpoints are connected and zero otherwise. In order to generate different types of networks, the choice of numbers for first- and final-stage products and the predecessor interconnection of stockpoints must be assigned with some priority mechanism. Therefore, a function $f(upper, type)$ is introduced that generates an integer random number from the interval $1 \leq X \leq upper$. The parameter $type$ represents a priority category from the set $\{large, medium, small\}$. The respective probabilities are

$$medium: \ Prob\{X = x\} = \frac{1}{upper},$$

$$large: \quad Prob\{X = x\} = \frac{2x}{upper(upper + 1)},$$

$$small: \quad Prob\{X = x\} = \frac{2(upper + 1 - x)}{upper(upper + 1)}.$$

If the number of first- and final-stage stockpoints is chosen sequentially, the outcome of the system will generally differ. Therefore, an additional priority category indicates if the first- or final-stage stockpoint number is determined first. If the final stage choice has higher priority ($init = end$),

$$|E| = f(n - 1, type)$$
$$|A| = f(n - |E|, type)$$

whereas the reverse priority of the first-stage stockpoint number ($init = start$) yields

$$|A| = f(n - 1, type)$$
$$|E| = f(n - |A|, type).$$

In a second step, the connection between stockpoints must be generated. It is assumed that supply relations are characterized by the material requirements $a_{ij} \in \{0, 1\}$. The number of predecessors for intermediate and final-stage stockpoints are chosen from

$$|v(i)| = \quad f(i - 1, type) \quad \forall i \in P,$$
$$|v(i)| = f(n - |E|, type) \; \forall i \in E.$$

Predecessor relations are allocated to candidate stockpoints by a Bernoulli probability mechanism where the probability of assigning $a_{j,i} = 1$ equals the number of relations that remain to be assigned divided by the remaining candidates. This mechanism is sequentially applied with increasing stockpoint index. If this interconnection choice provides a network with a stockpoint $i \in A \cup P$ without any successor, one end-item stockpoint successor is randomly selected with uniform distribution.

Using the above methodology, in total 50 network instances are generated with parameters ($init$, type($|A|$), type($|v(i)|$), type($|E|$)). 10 instances each are chosen out of the following 5 categories.

- "convergent"-logic: (start, large, large, small)
- "divergent"-logic: (end, small, small, large)
- "serial"-logic: (start, small, small, small)
- "mixed"-logic: (start, large, small, large)
- "equal"-logic: (end, medium, medium, medium)

The determination of demand, processing time, service level size, and holding cost data is addressed in the following.

Demand Data. Demands are independently identically distributed following a normal distribution. The expected value μ_i is chosen from the set $\{10, 20, ..., 100\}$ with $Prob\{X = \mu_i\} = 1/10$. Coefficients of variation v_i are uniformly chosen from the set $\{0.1, 0.2, 0.3, 0.4, 0.5\}$. Correlation between final-product demands is neglected.

Processing Times. Processing times are chosen from the set $\{1, ..., 5\}$ with equal probability. For final-stage points, the review period is added. It is assumed that the processing times of different stockpoints are independent and can, therefore, be chosen independently.

Service Level Size. All service levels are of the α-type. They are chosen from the set $\{0.90, 0.95, 0.99\}$ with uniform distribution for all stockpoints $i = 1, ..., n$.

Holding Cost. For the generation of holding costs, the principle of increasing holding cost due to value added within manufacturing is followed. Therefore, the cumulative holding cost of all direct predecessors is determined and a value chosen from the set $\delta_i \in \{1, ..., 5\}$ is added.

$$h_i = \sum_{j \in v(i)} h_j + \delta_i \qquad , Prob(X = \delta_i) = \tfrac{1}{5}$$

4.5.2 Definition of Performance Measures

For a comparison of algorithms and parameter settings, it is necessary to define performance measures. The measures used in the following concern solution quality on the one hand and computation time required on the other hand. Besides these two classes of criteria which represent quantitative measures, qualitative indicators as implementability or understandability are difficult to evaluate and therefore omitted.

C_i^* cost value of the optimal solution to problem instance i

C_i^m cost value of the solution to problem instance i determined with method m

N number of problem instances

The solution quality is analyzed with respect to three submeasures.

(1) Average number of optimal solutions found by method m

$$\#opt^m = \frac{1}{N} \cdot \sum_{i=1}^{N} 1_{[C_i^m = C_i^*]}$$

(2) Average cost deviation from the optimal solution (in %)

$$\overline{\Delta}^m = \frac{1}{N} \cdot \sum_{i=1}^{N} \frac{C_i^m - C_i^*}{C_i^*} \cdot 100$$

(3) Maximum deviation from the optimal solution (in %)

$$\Delta^m_{max} = \max_{i=1,...,N} \left\{ \frac{C^m_i - C^*_i}{C^*_i} \right\} \cdot 100$$

The computation time required to obtain the solution to the problem instances by method m is evaluated through the average and the maximum computation time in CPU seconds on a Pentium 90 MHz PC.

(1) Average computation time

$$\overline{CPU} = \frac{1}{N} \cdot \sum_{i=1}^{N} CPU(i)$$

(2) Maximum computation time

$$CPU_{max} = \max_{i=1,...,N} \{CPU(i)\}$$

4.5.3 Numerical Results

The optimal solutions to all problem instances have been determined by Enumeration. In a first step, the quality of the end-item safety stock placement strategy and the improvement potential of simple descent and steepest descent methods is reported in Table 4.9. Neighbor solutions were generated in a fixed order, starting with the stockpoint with lowest index.

Table 4.9. Performance of starting solution and local search

	# opt	$\Delta \overline{C}$	ΔC_{max}	$\overline{CPU}$	CPU_{max}
starting solution	4	7.9	34.0	0.0	0.0
steepest descent	19	2.4	17.8	0.7	1.8
descent	16	2.6	14.3	0.5	1.0

The starting solution yields an average cost gap (compared to the optimal strategic safety stock allocation solution) of 8% whereas the maximum gap was 34% . Specifically, the average cost gap is considerably reduced by both descent methods to about 2.5% whereas the worst case performance yields deviations of 18% and 14% respectively.

The Simulated Annealing algorithm was tested with an initial temperature choice following method (1b) and a stopping criterion (4c). A default parameter set for $P_1, ..., P_4$ was set to $P_1 = 1.0$, $P_2 = 0.90$, $P_3 = 1.0$, and $P_4 = 3.0$. Each parameter is varied with the other three remaining at their respective default value. The results are shown in Tables 4.10-4.13. The results indicate that under an appropriate setting of generic parameters, a solution

Table 4.10. Simulated Annealing: Initial temperature

P_1	# opt	$\overline{\Delta}$	Δ_{max}	$\overline{CPU}$	CPU_{max}
0.2	38	0.93	14.34	16.10	21.25
0.4	36	0.69	7.32	16.01	20.81
0.6	38	0.55	7.32	16.01	20.48
0.8	38	0.62	7.32	15.99	20.65
1.0	36	0.70	7.32	15.99	20.76
1.2	38	0.64	8.88	15.97	20.21
1.4	34	1.00	11.71	15.94	20.43
1.6	40	0.35	5.16	16.00	20.59

Table 4.11. Simulated Annealing: Temperature function

P_2	# opt	$\overline{\Delta}$	Δ_{max}	$\overline{CPU}$	CPU_{max}
0.50	35	1.10	11.78	15.99	21.09
0.60	33	1.05	7.32	15.97	20.98
0.70	36	0.82	7.32	16.11	20.92
0.80	36	0.70	7.32	16.02	22.08
0.85	35	0.87	12.02	15.98	21.47
0.90	40	0.51	7.32	16.06	21.14
0.95	41	0.35	5.16	16.00	20.54
0.99	35	0.50	5.16	15.99	21.31

Table 4.12. Simulated Annealing: Neighbors at each temperature

P_3	# opt	$\overline{\Delta C}$	ΔC_{max}	$\overline{CPU}$	CPU_{max}
0.2	33	0.9	8.9	4.4	5.9
0.4	33	0.9	8.9	7.4	9.3
0.6	38	0.7	8.9	10.4	15.1
0.8	35	0.7	5.2	13.5	21.7
1.0	40	0.5	7.3	16.4	20.9
1.5	42	0.3	5.2	24.2	30.9
2.0	40	0.5	5.2	31.2	40.8
2.5	41	0.4	5.2	39.5	52.0
3.0	42	0.3	5.2	46.5	60.3

performance of 80% optimal solutions, an average cost gap of 0.3% , and a worst case gap of 5% can be realized.

The Threshold Accepting algorithm is implemented with the same choices and parameter default values as the Simulated Annealing method. The results in Tables 4.14-4.17 show a similar performance as obtained for Simulated Annealing.

The Tabu Search algorithm is implemented with default parameters $P_1 = 3.0$ for the stopping criterion and $P_2 = 0.50$ for the list of forbidden moves. The results reported in Tables 4.18-4.19 are obtained with the

Table 4.13. Simulated Annealing: Temperature changes

P_4	# opt	$\overline{\Delta}$	Δ_{max}	$\overline{CPU}$	CPU_{max}
1	36	0.65	5.51	5.47	6.97
2	35	0.61	5.16	10.74	13.56
3	37	0.61	7.32	16.00	20.43
4	40	0.31	5.16	21.39	28.17
5	38	0.58	7.32	26.56	34.87

Table 4.14. Threshold Accepting: Initial threshold

P_1	# opt	$\overline{\Delta}$	Δ_{max}	$\overline{CPU}$	CPU_{max}
0.2	37	0.95	14.34	16.14	21.42
0.4	36	0.92	13.52	16.13	21.42
0.6	38	0.64	7.32	16.16	21.03
0.8	39	0.53	5.16	16.10	20.98
1.0	37	0.69	7.32	16.16	20.87
1.2	38	0.49	5.16	16.26	23.83
1.4	38	0.55	6.65	16.23	21.20
1.6	37	0.59	7.32	16.18	20.59

Table 4.15. Threshold Accepting: Threshold lowering function

P_2	# opt	$\overline{\Delta}$	Δ_{max}	$\overline{CPU}$	CPU_{max}
0.50	33	0.91	8.88	16.00	21.42
0.60	34	1.08	13.35	16.06	21.25
0.70	35	0.91	8.88	16.02	21.09
0.80	38	0.53	5.16	16.08	20.87
0.85	38	0.58	7.32	16.10	21.03
0.90	42	0.31	5.16	16.15	20.48
0.95	39	0.37	5.16	16.10	20.26
0.99	35	0.57	5.16	16.04	21.69

variant that uses the discussed aspiration level criterion to accept a forbidden solution if it provides a better solution than the best configuration found up to that iteration. Compared to Simulated Annealing and Threshold Accepting, a dominance of the Tabu Search approach with respect to the number of optimal solutions, average and maximum cost gap, as well as in required computation times appears. Nevertheless, from an application point of view, all methods provide an accuracy with deviations that lie below the level of data accuracy. Furthermore, the implemented algorithms represent basic variants that can be improved by more elaborate and elegant search features but the realized additional improvement will only be small, whereas the basic implementations can already be recommended for solving the safety stock allocation problem in general network supply chains.

Table 4.16. Threshold Accepting: Evaluated neighbors at each threshold value

P_3	# opt	$\overline{\Delta}$	Δ_{max}	$\overline{CPU}$	CPU_{max}
0.2	33	1.00	11.78	4.29	5.54
0.4	35	0.72	5.16	7.25	9.22
0.6	36	0.65	8.88	10.06	12.96
0.8	39	0.56	8.88	13.09	16.53
1.0	40	0.40	5.16	16.06	20.87
1.5	38	0.45	5.16	23.83	30.04
2.0	40	0.37	5.16	30.88	41.52
2.5	39	0.46	5.16	38.58	49.92
3.0	42	0.33	5.16	45.82	59.70

Table 4.17. Threshold Accepting: Threshold lowering number

P_4	# opt	$\overline{\Delta}$	Δ_{max}	$\overline{CPU}$	CPU_{max}
1	31	0.83	6.83	5.43	7.03
2	41	0.37	5.16	10.72	13.62
3	38	0.60	7.32	15.92	21.14
4	38	0.58	5.16	21.39	27.24
5	38	0.49	5.16	26.54	34.21

Table 4.18. Tabu Search: Tabu list length

P_2	# opt	$\overline{\Delta C}$	ΔC_{max}	$\overline{CPU}$	CPU_{max}
0.1	24	1.8	14.3	10.8	17.5
0.2	28	1.5	14.3	10.9	17.6
0.3	33	0.9	7.3	11.1	17.6
0.4	38	0.5	4.8	11.2	17.6
0.5	40	0.4	4.8	11.2	17.6
0.6	41	0.2	4.4	11.3	17.6
0.7	43	0.2	3.4	11.4	18.2
0.8	44	0.1	2.7	11.4	18.5
0.9	45	0.1	2.7	11.5	19.4

Table 4.19. Tabu search: Number of iterations

P_1	# opt	$\overline{\Delta C}$	ΔC_{max}	$\overline{CPU}$	CPU_{max}
1	35	0.8	11.7	3.7	5.8
2	40	0.4	4.8	7.5	11.7
3	40	0.4	4.8	11.2	17.6
4	41	0.3	4.8	15.0	23.5
5	41	0.3	4.8	18.8	29.5

4.6 Cyclic Systems with External and Internal Returns

The model introduced in 4.4 assumes an acyclic network. In this section, external and internal forward and backward return flows are integrated. First, additional assumptions concerning these flows are summarized before the extended model is analyzed. Finally, the optimization problem is presented and solution method adjustments are suggested.

4.6.1 Model Formulation

Products are randomly returned by external customers to stockpoints $i \in F^E$. These stockpoints can be either at the final stage of the supply chain (i.e. merchandise returns) or at an internal stage (i.e. recovery returns). It is assumed that returns R_i are normally distributed with mean $\mu_{i,bp}$ and standard deviation $\sigma_{i,bp}$. Correlation of return quantities in time as well as correlation between demands and returns are neglected. A justification for assuming independence is that, in most cases, the operating time will be smaller than the correlation delay between demands and returns. Internal returns are induced by processes. In extension to the assumption of a single product outcome of each process, some processes yield by-products that are recovered and then reused as material substitutes in other processes $i \in F^I$. Depending on the location of the reuse stockpoint, backward and forward reuse are distinguished. Backward reuse occurs if by-products of j are reused at a stockpoint $i \in V(j)$, or in other words, the recovered material revisits a process. If recovered by-products are reused in a separate part of the supply chain, for instance in a different product line, this implies a forward reuse operation. Therefore, a forward reuse arc can never end in a stockpoint $i \in N(j)$. The by-product quantity that is created by process j and reused as a substitute for product i is assumed to be a deterministic fraction $a_{j,i}^{bp}$ of the desired main product outcome quantity at $j = bp(i)$. To keep the model simple, it is further assumed that each by-product of j has a single destination stockpoint, which implies that no allocation decisions concerning the by-products are necessary and that each stockpoint at most reuses material from one (external or internal) return source ($F^E \cap F^I = \emptyset$). This implies that, together with the regular replenishment source, at most a two supply mode consideration is necessary. Nevertheless, an extension to multiple (> 2) modes is straightforward. Returns and by-products are neither disposed nor kept in inventory. All returns and by-products are immediately recovered with a known time requirement λ_i^{bp} and then kept in stock together with the regularly manufactured products at the same holding cost rate. To account for recovered returns that serve as substitutes for regularly produced items, replenishment policies of stockpoints $i \in F^E \cup F^I$ must be adjusted. It is assumed that each stockpoint operates with a single base-stock level. Return quantities of period t are subtracted from the regular replenishment order in the same time period.

In case of external returns and forward by-product reuse, returns can exceed the demand driven requirements so that excess returns are subtracted from future regular replenishment orders. Note that for backward reuse operations, the cyclic dependency that the by-product outcome can only be a fraction of the input, avoids this effect. Given the characteristics of customer demand $D_{i,t}$ $\forall i \in E$ and external returns $R_{i,t}$ $\forall i \in F^E$ and the reuse relationships, internal demand distributions must be adjusted with respect to the netting of gross requirements $D_{i,t}^{gross}$ with return quantities. The regular mode replenishment quantity $Q_{i,t}$ is given by the difference of direct gross requirements and direct returns.

$$D_{i,t}^{gross} = D_{i,t} \qquad \forall i \in E$$

$$D_{i,t}^{gross} = \sum_{j \in n(i)} a_{i,j} \cdot Q_{j,t} \; \forall i \in A \cup P$$

$$Q_{i,t} = D_{i,t}^{gross} - R_{i,t} \; \forall i \in A \cup P \cup E$$

The last equation neglects the case where returns exceed gross requirements. From the long run strategic perspective, R will be relatively small in comparison to D so that compensation from netting these excess returns with the next positive regular replenishment is achieved fast. The assumption of an immediate netting with pipeline orders, instead of a later netting, even overestimates real safety stock requirements. In case of external returns, $R_{i,t}$ denotes an independent random variable whereas $R_{i,t}$ for $i \in F^I$ denotes a random variable that depends on the replenishment quantity $Q_{bp(i),t}$ of the by-product generating process $j = bp(i)$.

$$R_{i,t} = a_{bp(i),i}^{bp} \cdot Q_{bp(i),t}$$

The system of linear equations can be solved for the total requirements coefficients $g_{i,j}$. Then, given the external demand and return vector, internal net demand is

$$Q_{i,t} = \sum_{j \in E \cup F^E} g_{i,j} \cdot (D_{j,t} - R_{j,t}). \tag{4.19}$$

Ordering and replenishment coordination for each stockpoint concerns at most two supply modes, the regular mode and the return reuse mode. Using the regular replenishment mode, net requirements $Q_{i,t}$ are replenished with lead time $L_i = \max_{j \in v(i)}\{ST_j\} + \lambda_i$. The reuse mode provides recovered returns $R_{i,t}$ after $L_i^{bp} = \lambda_i^{bp}$ periods for external and after $L_i^{bp} = L_{bp(i)} + \lambda_i^{bp}$ periods for internal returns. Because the service times are strategic decision variables, the total delay in internal recovery activities is variable as well by $L_{bp(i)}$. The main replenishment mode is given by regular ordering. Therefore, the service time ST_i of stockpoint i is bounded by the regular replenishment lead time L_i ($ST_i \leq L_i$). The coverage time $T_i = L_i - ST_i$ has to be adjusted with respect to the two mode replenishment with (in general) offsetting lead times. The total replenishment lead time for gross requirements of a stock-point is given by the larger one of the two lead times.

$$\tilde{T}_i(ST_i, L_i, L_i^{bp}) = \max\{L_i; L_i^{bp}\} - ST_i$$

Therefore, the first period that may be completely influenced is $t + \tilde{T}_i + ST_i$. In order to derive a safety stock size to cover against the uncertainty within the $\tilde{T}_i$ periods that accounts for offsetting lead times, that is that one of the two modes provides an earlier availability of material, the net demand characteristic over the time interval $[t+ST_i; t+\tilde{T}_i+ST_i]$ is examined. Depending on the values of ST_i, L_i, L_i^{bp} three cases are distinguished.

Case 1: $ST_i \leq L_i \leq L_i^{bp}$

If the service time is smaller than both replenishment lead times, safety stock coverage is required for both supply modes over gross requirements variations of $L_i^{bp} - ST_i$ time periods ($[t + ST_i; t + L_i^{bp}]$). Within this time interval, net regular replenishment orders over $L_i^{bp} - L_i$ time periods ($[t + L_i; t + L_i^{bp}]$) are received and can be subtracted.

$$D_i^{net}[L_i, L_i^{bp}, ST_i] = D_i^{gross}[t + ST_i; t + L_i^{bp}] - Q_i[t + L_i; t + L_i^{bp}]$$
$$= D_i^{gross}[t + ST_i; t + L_i] + R_i[t + L_i; t + L_i^{bp}]$$

Case 2: $ST_i \leq L_i^{bp} \leq L_i$

Safety stock coverage is required for both supply modes. Gross requirements are taken into account over $L_i - ST_i$ time periods ($[t + ST_i; t + L_i]$). In this case, reuse return flows over $L_i - L_i^{bp}$ time periods ($[t + L_i^{bp}; t + L_i]$) are received within the coverage time interval.

$$D_i^{net}[L_i, L_i^{bp}, ST_i] = D_i^{gross}[t + ST_i; t + L_i] - R_i[t + L_i^{bp}; t + L_i]$$
$$= D_i^{gross}[t + ST_i; t + L_i^{bp}] + Q_i[t + L_i^{bp}; t + L_i]$$

Case 3: $L_i^{bp} \leq ST_i \leq L_i$

In the third case where the return recovery processing time is smaller than the service time, safety stock requirements exist only for the regular replenishment mode. Earlier reuse product availability can be exploited for buffering purposes. Gross requirements over $L_i - ST_i$ time periods ($[t+ST_i; t+L_i]$) are netted with received return flows over $L_i - L_i^{bp}$ time periods ($[t+L_i^{bp}; t+L_i]$). This can equivalently be expressed by net regular replenishment orders over $L_i - ST_i$ time periods less available reuse material over $ST_i - L_i^{bp}$ periods.

$$D_i^{net}[L_i, L_i^{bp}, ST_i] = D_i^{gross}[t + ST_i; t + L_i] - R_i[t + L_i^{bp}; t + L_i]$$
$$= Q_i[t + ST_i; t + L_i] - R_i[t + L_i^{bp}; t + ST_i]$$

The safety stock quantity requirement that follows from net demand over $\tilde{T}_i$ can be expressed by the standard deviation of net demand within this time interval. In the safety stock formula, $\sigma_i \cdot \sqrt{T_i}$ has to be replaced by $\sigma_i^{net}(L_i, L_i^{bp}, ST_i) = \sqrt{Var(D_i^{net}(L_i, L_i^{bp}, ST_i))}$. When computing these standard deviations, care has to be given concerning the evaluation of variances for demand and return expressions of overlapping time periods.

4.6.2 Solution Principle

By incorporating the adjustments with respect to external and internal returns into the optimization problem formulation, the extensions only influence the objective function but not the feasible region. Therefore, the extreme points of the solution set are identical to the ones discussed for the basic model. Additional complexity arises from the separation into three net demand cases. Because similar concavity arguments hold for each of the three regions, the separating values must be taken into account additionally. This leads to an extended extreme point representation of relevant safety stock allocation solutions which implies a kind of synchronization of service times and reuse mode.

(1) $ST_i = L_i^{bp}$

For this type of synchronization, the service time quoted to successors is identical to the reuse lead time. These additional solutions are only possible for stockpoints i with external return or forward by-product reuse.

(2) $L_i = L_i^{bp}$

In this case, coverage and service times are chosen in a way that both replenishment lead times become identical and fully synchronized.

A first approach to find a solution to the extended optimization problem is to neglect the additional synchronization options that add a large amount on computational complexity and to restrict to bang-bang policies, that is to cover nothing or the entire regular replenishment lead time. In this case, the combinatorial solution principle applies except for safety stock size adjustments with respect to σ_i^{net} in the evaluation of a given safety stock allocation representation $\underline{b}$. As a second approach, a generalized encoding scheme that accounts for additional synchronization options can be implemented. For the safety stock allocation representation vector, $b(i) = 2$ implies that stockpoint i holds safety stock in order to guarantee a service time that is identical to the reuse lead time. Every predecessor i of a stockpoint j that faces a return input can synchronize its service time. This is indicated by $b(i) = -j$. Because of logical interaction between synchronization and extreme point properties, not all combinations of the above representations are relevant. In order to be a synchronization solution, no successor l on a path between i and j can cover the entire replenishment lead time ($b(l) = 1$) or synchronize for the same destination reuse node $b(l) = -j$. With respect to the resulting coverage time determination, lead time calculations are adjusted for the additional values of $b(i)$.

$$L_i = \lambda_i + \max_{j \in v(i)} \{\tilde{m}_j\}$$

The values $\tilde{m}_j$ represent the service time outcome of predecessor j depending on $b(j)$ and are defined as follows.

$$b(i) = 0 : \tilde{m}_i = L_i$$
$$b(i) = 1 : \tilde{m}_i = 0$$
$$b(i) = 2 : \tilde{m}_i = \begin{cases} \lambda_i^{bp} & i \in F^E \\ L_{bp(i)} + \lambda_i^{bp} & i \in F^I \end{cases}$$
$$b(i) = -j :$$

$$\tilde{m}_i = \begin{cases} (L_j^{bp} - \max\limits_{w \in W(i,j)} \{\sum\limits_{m \in w \backslash \{i\}} \lambda_m\})^+ & (j \in F^E) \\ & \vee (j \in F^I \wedge N(i) \cap bp(j) = \emptyset) \\ (L_j - \max\limits_{w \in W(i,j)} \{\sum\limits_{m \in w \backslash \{i\}} \lambda_m\} - \lambda_j^{bp})^+ & j \in F^I \wedge N(i) \cap bp(j) \neq \emptyset \end{cases}$$

If $b(i) = 0$, no coverage is planned and the entire (regular) replenishment lead time has to be covered by all successors. If $b(i) = 1$, all demand uncertainty is covered and no coverage obligations are postponed to succeeding stockpoints. In case of $b(i) = 2$, the reuse mode lead time defines the service time and the difference (if positive) between reuse mode and regular mode time is covered. If i offers a service time in order to synchronize both mode lead times of j, the calculation depends on whether (1) j is a stockpoint with external returns or j is an internal reuse stockpoint and i lies on the regular replenishment path of j or (2) i lies on the by-product supply path to j. In case (1), the service time is chosen in a way that service time plus the maximum cumulative time of succeeding processes equals the reuse mode lead time of j. In case (2), the sum of service time of i, maximum cumulative processing time, and recovery time equals the regular mode replenishment time. Note that the required L_i values are known in the backward reuse case and can be obtained in the forward reuse case due to the restriction that forward reuse is only possible in a different product line. For implementation of this extended solution representation within a local search framework, the neighborhood definition has to undergo some adjustments. Where the representation presented in Section 2 offers only a single transition $(1 - b(i))$ for each stockpoint, the extended version additionally offers transitions to and from synchronization solutions $b(i) = 2$ and $b(i) = -j$.

The impact of incorporating returns into strategic safety stock placement considerations is illustrated by an example. The original network is a four-stage assembly supply chain with two purchasing processes (1,2) and a single final product (5). The dotted line represents the incorporation of an internal backward reuse arc. Besides the desired manufacturing product 4, the corresponding process yields a by-product outcome that can be reused at stockpoint 2.

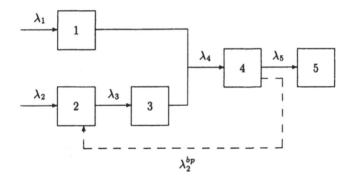

Fig. 4.4. Example network with one return flow arc

Net demand analysis for the second stockpoint with a regular supply mode $(L_2 = \lambda_2)$ and a reuse mode $(L_2^{bp} = \max\{ST_1; ST_3\} + \lambda_4 + \lambda_2^{bp})$ gives the following result:

Case 1: $ST_2 \leq L_2 \leq L_2^{bp}$

$$D_2^{net}[L_2, L_2^{bp}, ST_2] = a_{2,5} \cdot D_5[t + ST_2; t + L_2]$$
$$+ a_{4,2}^{bp} \cdot a_{4,5} \cdot D_5[t + L_2; t + L_2^{bp}]$$

Case 2: $ST_2 \leq L_2^{bp} \leq L_2$

$$D_2^{net}[L_2, L_2^{bp}, ST_2] = a_{2,5} \cdot D_5[t + ST_2; t + L_2^{bp}]$$
$$+ (a_{2,5} - a_{4,2}^{bp}) \cdot D_5[t + L_2^{bp}; t + L_2]$$

Case 3: $L_2^{bp} \leq ST_2 \leq L_2$

$$D_2^{net}[L_2, L_2^{bp}, ST_2] = (a_{2,5} - a_{4,2}^{bp}) \cdot D_5[t + ST_2; t + L_2]$$
$$- a_{4,2}^{bp} \cdot D_5[t + L_2^{bp}; t + ST_2]$$

with $a_{2,5} = a_{2,3} \cdot a_{3,4} \cdot a_{4,5}$. In Example 1 it is shown that the optimal solution can be of the synchronization type. Example 2 presents the natural case where the incorporation of reuse creates additional inventory. Example 3 shows that the system with reuse may require less safety stocks than the original system.
Example 1: Optimality of a synchronization solution
Final product demand is normally distributed with $\mu_5 = 10, \sigma_5 = 3$. All direct material requirements coefficients are $a_{i,j} = 1$ except for $a_{3,4} = 1.5$ and the by-product coefficient $a_{4,2}^{bp} = 0.4$. The reuse processing time λ_2^{bp} is zero. The coefficients for stockpoints $i = 1, 2, 3, 4, 5$ are processing times λ_i (2,4,1,1,2), service levels α_i (0.99,0.99,0.90,0.90,0.95), and value added holding costs h_i (3,3,3.5,8.5,11.5).

The optimization results are summarized in Table 4.20. Besides the internal demand parameters, the optimal safety stock allocation with corresponding coverage time and safety stock value are given. The traditional system, without returns, yields a safety stock policy with lower costs than the policy under return reuse. For the optimal solution of the new system with reuse

Table 4.20. Optimization results for Example 1

	old system					new system				
	1	2	3	4	5	1	2	3	4	5
μ_i	10	15	15	10	10	10	11	15	10	10
σ_i	3	4.5	4.5	3	3	3	3.3	4.5	3	3
$b(i)$	0	0	0	1	1	0	0	2	0	1
T_i	0	0	0	6	2	0	0	2	0	6
SS_i	0	0	0	9.42	6.98	0	0	8.16	0	12.09
C			160.33					167.58		

flow, $L_2 = L_2^{bp} = 4$ holds and additional inventory at stockpoint 2 is avoided. To illustrate the result in more detail, this solution is compared to the best bang-bang solution for the new system and to the solution with the same allocation pattern as for the traditional solution. The best bang-bang solution is given by $\underline{b} = (0, 0, 1, 0, 1)$, $\underline{T} = (0, 0, 3, 0, 5)$ and $\underline{SS} = (0, 2.79, 9.99, 0, 11.04)$ with $C = 170.26$. The fact that the two mode lead times of stockpoint 2 $(L_2 = 4, L_2^{bp} = 3)$ are not synchronized creates additional inventory at this stockpoint. The same effect occurs when using the optimal allocation pattern of the traditional system. The only difference is that additional inventory of 3.95 units is created at stockpoint 2 ($C = 172.18$).

Example 2: Optimal safety stock solution of the system without reuse arc dominates system with reuse

The probability distribution parameters are again $\mu_5 = 10, \sigma_5 = 3$. Direct material requirements are given by $a_{i,j} = 1$, except for $a_{3,4} = 1.5$ and $a_{4,2}^{bp} = 0.5$. Return processing requires $\lambda_2^{bp} = 1$ period. Processing times for regular processes λ_i are (3,1,1,1,2), service levels α_i (0.95,0.95,0.95,0.95,0.95), and holding cost rates h_i (1,1,2,4,4.5). When implementing the optimal allocation pattern of the traditional assembly system, additional inventory of 4.94 ($C = 59.34$) is created at the second stockpoint without opening the opportunity to lower safety inventory at other stockpoints. The optimal recovery system solution cannot avoid this effect but additional inventory can be decreased by holding safety stock at stockpoint 1 which provides an additional safety stock reduction at the final stage.

Example 3: Optimal safety stock solution of reuse system dominates the traditional system

Customer demand, material requirements coefficients, by-product outcome, return processing time and service levels are the same given for Example 2.

Table 4.21. Optimization results for Example 2

	old system					new system				
	1	2	3	4	5	1	2	3	4	5
μ_i	10	15	15	10	10	10	10	15	10	10
σ_i	3	4.5	4.5	3	3	3	3	4.5	3	3
$b(i)$	0	0	0	0	1	1	0	0	0	1
T_i	0	0	0	0	6	1	0	0	0	5
SS_i	0	0	0	0	12.24	4.94	4.27	0	0	11.04
C	54.04					58.87				

Processing times λ_i are (1,4,1,1,2) and holding cost rates h_i (2,1,4,8,12). The

Table 4.22. Optimization results for Example 3

	old system					new system				
	1	2	3	4	5	1	2	3	4	5
μ_i	10	15	15	10	10	10	10	15	10	10
σ_i	3	4.5	4.5	3	3	3	3	4.5	3	3
$b(i)$	0	1	0	0	1	0	1	0	0	1
T_i	0	4	0	0	4	0	4	0	0	4
SS_i	0	14.81	0	0	9.87	0	13.74	0	0	9.87
C	133.26					132.20				

effect that the return recovery system is able to operate with less inventory than the corresponding system without return integration is achieved by the coordination of material flows in a way that the return material becomes available ($L_i^{bp} = 3$) before regular replenishments ($L_i = 4$) and this stock can be used for reducing safety inventory.

4.7 Extensions

The solution approach presented for general multi-echelon networks represents the straightforward extension of the SIMPSON approach developed for serial systems. Several extensions to requirements of practical applications have been discussed for serial systems and apply for general supply chains with some slight modifications concerning safety stock sizing problems or, on the other hand, coverage allocation methodology.

The incorporation of lost internal demands in situations of emergency supplies necessitates an adjustment in required safety stock sizes. A general network supply chain that operates under γ- instead of α-service levels requires safety stock adjustments with respect to the use of a different safety factor. Additionally, problem reduction properties and potential coverage reductions with respect to τ-values might occur and result in an additional coverage logic adjustment. The same argument applies for stockpoints that replenish material requirements in batches because of the similarity to the γ-service level case.

An extension to stochastic processing times requires a different type of coverage logic adjustment. A safety time increment to the expected processing times has to be covered, in addition to the demand uncertainty driven buffer stocks. If service times are differentiated with respect to direct successors or, as a further extension, with respect to corresponding final-product requirements, the two model extensions introduced for divergent systems apply. In addition, these extensions are easily incorporated into the heuristic local search solution procedures by introducing double indexed binary coverage variables.

5. Concluding Remarks

The main characteristic of the presented approach is the service time logic that every stockpoint supplies its successors after a predetermined time span. The strategic decision upon these inventory control policy parameters is connected to the safety stock planning approach. Reasonable demand variability is absorbed by these strategic buffers, whereas extraordinary emergency situations are left to operating flexibility. Its applicability to general network supply chains, the availability of several heuristic solution methods with satisfactory accuracy, and further extendability of the underlying logic makes this approach favorable for utilization as a standardized tool for safety stock planning.

Though the presented approach aims to provide general guidelines for safety stock planning, it will be better suited for some model environments than others. The service orientation of the approach is better suited for make-to-stock environments than to the due date directed make-to-order strategy. The applied periodic review policies that assume a demand distribution for a base period are mostly applied for fast moving (e.g. consumable) products, whereas continuous review policies which assume stochastic processes better fit for slow moving, expensive (e.g. spare parts) product applications. Finally, it better applies to old, mature product life cycle phase products with large and predictable demand, which justifies the use of theoretical demand distributions, whereas new products with unstable or even unknown demand parameters and distributions are better handled by strategies like quick and accurate response.

In addition to the solution of the strategic safety stock planning problem, the presented models can serve as a building block for model based decisions within other logistics management or business areas and for what-if analysis. For different design decisions for production/inventory networks in a stochastic environment, the materials coordination cost and service trade-offs can be roughly evaluated. The material logistics consequences of delivery contracts offered by sales forces to customers as well as supply conditions of several suppliers can be analyzed and serve as a basis for negotiations between different organizational units of an enterprise.

Besides the extensions already discussed or mentioned, some aspects offer a potential for further research. The increasing importance of capacitated production systems already influenced the research in scheduling and lot-sizing models. For the multi-stage safety stock planning problem, only a few publications are available.[1] In this area, the analysis of capacity effects on lead time variability by queuing models appears fruitful. A more detailed incorporation of operative flexibility consequences into the strategic safety stock model represents a second important research topic. Instead of coupling operating flexibility to maximum reasonable demand and therefore to internal service level constraints, more detailed emergency supply, process lead time flexibility, or product scheduling models can provide an improved anticipation of the respective consequences within the strategic level allocation model.

---

[1] GLASSERMAN, P. [1997].

Appendix

A. Concavity Properties

In this section, some properties of safety and on-hand stock will be derived under normally distributed demands. The objective is to show that stock levels increase with larger coverage time in a concave way.

A.1 Order-up-to-Policy

A.1.1 Safety Stock

$$SS = \int_{-\infty}^{\infty} (S - x)\varphi_T(x)dx = S - T \cdot \mu = k\sigma\sqrt{T} \qquad \text{(A.1)}$$

Lemma A.1.1. *Under an α-service level constraint, the safety stock is a concave function of the coverage time.*

Proof

$$\frac{dSS}{dT} = \frac{k\sigma}{2\sqrt{T}} > 0$$

$$\frac{d^2SS}{dT^2} = -\frac{k\sigma}{4T\sqrt{T}} < 0$$

$\square$

Lemma A.1.2. *Under a γ-service level constraint, the safety stock is a concave function of the coverage time if the safety stock is restricted to non-negative values.*

Proof In order to prove the lemma, $\frac{d^2C(T)}{dT^2} \leq 0$ with $C(T) = h \cdot \sigma \cdot k(T) \cdot \sqrt{T}$, $k(T)$ given as outlined in Section 2.3, has to be verified. It turns out that

$$\frac{dC(T)}{dT} = h\sigma \left(\frac{dk(T)}{dT} \cdot \sqrt{T} + k(T) \cdot \frac{1}{2\sqrt{T}} \right) > 0$$

which proves the monotonicity of $C(T)$, and further

$$\frac{d^2C(T)}{dT^2} = h\sigma \left(\frac{d^2k(T)}{dT^2} \cdot \sqrt{T} + \frac{dk(T)}{dT} \cdot \frac{1}{\sqrt{T}} - \frac{1}{4} \cdot \frac{k(T)}{T \cdot \sqrt{T}} \right)$$

$$= \frac{h\sigma}{4T\sqrt{T}} \left(\frac{\phi(k(T)) \left((1-\gamma) \cdot \frac{\mu}{\sigma} \right)^2}{-T \cdot (\Phi(k(T)) - 1)^3} + \frac{T^{-\frac{1}{2}}(1-\gamma) \cdot \frac{\mu}{\sigma}}{\Phi(k(T)) - 1} - k(T) \right)$$

where $\frac{d^2k(T)}{dT^2}$ and $\frac{dk(T)}{dT} > 0$ are obtained by implicit differentiation. Because $G(k) = (1-\gamma)\frac{\mu}{\sigma \cdot \sqrt{T}}$ and $G(k) = \phi(k) + k \cdot [\Phi(k) - 1]$ for normally distributed demands, $G(k)^2 \leq \Phi(k) - 1^2$ represents the condition for concavity. Under $G(k) \geq 0 \; \forall k \geq 0$ this can be transformed into $G(k) \leq 1 - \Phi(k)$. Again substituting $G(k)$ from above leads to

$$\frac{\phi(k)}{k+1} + \Phi(k) \leq 1$$

which is valid for all $k \geq 0$ because the left-hand side is increasing in k and

$$\lim_{k \to \infty} \frac{\phi(k)}{k+1} + \Phi(k) = 1.$$

$\square$

A.1.2 On-Hand Stock

$$OH = \int_{-\infty}^{S} (S - x)\varphi_T(x)dx = S \cdot \Psi_T(S) - \int_{-\infty}^{S} x\varphi_T(x)dx \qquad \text{(A.2)}$$

After standardization this simplifies to

$$OH = \sigma\sqrt{T} \int_{-\infty}^{k} (k - x)\phi(x)dx$$

$$= \sigma\sqrt{T}(k\Phi(k) + \phi(k)).$$

Lemma A.1.3. *Under an α-service level constraint, the on-hand stock is a concave function of the coverage time.*

Proof

$$\frac{dOH}{dT} = \frac{\sigma}{2\sqrt{T}} (k\Phi(k) + \phi(k)) > 0$$

$$\frac{d^2OH}{dT^2} = -\frac{1}{4T\sqrt{T}} (k\Phi(k) + \phi(k)) < 0$$

$\square$

Lemma A.1.4. *Under a γ-service level constraint, the on-hand stock is a concave function of the coverage time if the safety stock is restricted to non-negative values.*

Proof

$$\frac{dOH}{dT} = \frac{\sigma}{2\sqrt{T}}(k\Phi(k) + \phi(k)) + \sigma\sqrt{T}\Phi(k)\frac{dk}{dT} > 0$$

$$\frac{d^2OH}{dT^2} = -\frac{\sigma}{4T\sqrt{T}}(k\Phi(k) + \phi(k)) + \frac{\sigma}{\sqrt{T}}\left(\frac{dk}{dT}\Phi(k)\right)$$

$$+ \sigma\sqrt{T}\left(\frac{d^2k}{dT^2}\Phi(k) + \phi(k)\left(\frac{dk}{dT}\right)^2\right)$$

$$= \Phi(k)\left(-k\frac{\sigma}{4T\sqrt{T}} + \frac{\sigma}{\sqrt{T}}\frac{dk}{dT} + \sigma\sqrt{T}\frac{d^2k}{dT^2}\right)$$

$$+ \phi(k)\left(\sigma\sqrt{T}\left(\frac{dk}{dT}\right)^2 - \frac{\sigma}{4T\sqrt{T}}\right)$$

The term within the first brackets is the same as analyzed for γ-service level constraints and the safety stock criterion. The term within the second brackets is not larger than zero under the same condition. Then,

$$\left(\frac{dk}{dT}\right)^2 \le \frac{1}{4T^2}, \quad (G(k))^2 \le (\Phi(k) - 1)^2.$$

□

A.2 Reorder-Point, Order-Level-Policy

Lemma A.2.1. *Under an α-service level constraint, the safety stock is a concave function of the coverage time.*

Because of the analogy of the order-up-to-level problem under γ-service level constraints and the batch ordering problem under α-service level constraints, the proof is the same given in the previous section.

Lemma A.2.2. *Under a γ-service level constraint, safety stock is a concave function of the coverage time if the safety stock is restricted to non-negative values.*

Proof The implicit function for safety factor determination is $I(k) = J(k) - c/T$ with $c = (2\overline{Q}\mu + \sigma^2 + \mu^2)/\sigma^2$. For the quadratic loss integral it holds

$$J(k) = (1 + k^2)(1 - \Phi(k) - k\phi(k)) = 1 - \Phi(k) - kG(k).$$

τ determined by setting $k = 0$ yields $J(0) = 1/2$ and

$$\tau = 2(1 - \gamma)\frac{2\overline{Q}\mu + \sigma^2 + \mu^2}{\sigma^2}.$$

Further, $\frac{dG(k)}{dk} = \Phi(k) - 1$, $\frac{d^2G(k)}{dk^2} = \phi(k)$, $\frac{dI(k)}{dk} = -2G(k)$, and $\frac{d^2I(k)}{dk^2} = -2(\Phi(k) - 1)$. Using implicit differentiation yields

$$\frac{dI(k)}{dT} = \frac{dI(k)}{dk}\frac{dk}{dT} + \frac{c}{T^2}.$$

Then, the first two derivatives of the safety factor k respect to the coverage time T are given by

$$\frac{dk}{dT} = \frac{c}{2T^2G(k)} > 0$$

$$\frac{d^2k}{dT^2} = \frac{-c}{T^3G(k)} + \frac{-c(\Phi(k) - 1)}{2T^2G(k)^2}\frac{dk}{dT}.$$

As already shown for lot-for-lot ordering, it must be proven that

$$\frac{d^2k}{dT^2}\sqrt{T} + \frac{dk}{dT}\frac{1}{\sqrt{T}} - \frac{k}{4T\sqrt{T}} \leq 0.$$

Inserting $I(k) = c/T$ and simplifying terms yields

$$-2I(k)G(k)^2 + (1 - \Phi(k))I(k)^2 - kG(k)^3 \leq 0.$$

For $k = 0$, $\frac{\pi-4}{8\pi} \leq 0$. For the limiting value $\lim_{k\to\infty}$ the equation holds as an equality and it remains to show that the left hand side of the equation is monotonously increasing. Simplyfying the first derivative yields $3G(k)^2 - I(k)^2 \geq 0$. Repeating the argument, we find $\frac{6-\pi}{4\pi} > 0$ for $k = 0$ and $\lim_{k\to\infty} = 0$. The left hand side of the equation is monotonously decreasing which can be observed from the first derivative which after simplification and substitution yields $2(\Phi(k) - 1) - 4kG(k) \leq 0$. $\square$

A.3 Solution Separation

In some models, concavity of the objective function over the entire solution set is invalid. Nevertheless, the objective function can be shown to be concave over convex subsets of the original feasible region. In this appendix, this reasoning is applied to the γ-service level constraint problem in divergent networks.

Let $z_i = \sum_{j \in V(i)\setminus\{i\}}(\lambda_j - T_j) \geq 0$ denote the sum of uncovered processing times up to stockpoint i (without processing time of i). Due to the successive character of the $\leq$ constraints in the divergent optimization problem, coverage times are restricted by $0 \leq T_i \leq \lambda_i + z_i$. Starting with final-stage stockpoints, two cases are distinguished. In the case $\lambda_i + z_i \geq \tau_i$, the part of the objective function directly depending on T_i is concave in the coverage time. In the opposite case, the objective function does not depend on T_i. Because the division into the two cases depends on z_i, critical values for z_i which separate the cases are derived. These values are given by $\delta(i) = \max\{0 \; ; \; \tau_i - \lambda_i\} \; \forall i \in E$ because the second case is impossible if $\lambda_i > \tau_i$.

The analysis for a non-final-stage stockpoint i requires a set H which contains all critical values obtained for direct successors. Let $\delta(i,1) > \delta(i,2) > \ldots > \delta(i,|H|)$ denote the values within this set in decreasing order. Depending on z_i, two cases have to be distinguished.

Case 1: $\lambda_i + z_i < \tau_i \Rightarrow T_i = \lambda_i + z_i$

Here, the coverage time is fixed and therefore, no downstream decisions are affected.

Case 2: $\lambda_i + z_i \geq \tau_i \Rightarrow T_i \geq \tau_i$

In this case, the costs that directly depend on T_i are concave. To be able to identify the relevant values for succeeding stockpoints, it is necessary to divide the solution set concerning i ($\tau_i \leq T_i \leq \lambda_i + z_i$) into $|H| + 1$ disjunct subsets. The relevant coverage times of successors j depend on $z_j = z_i + \lambda_i - T_i$ and on the relation to the critical values in H.

$$\text{Subset 1:} \qquad \lambda_i + z_i - T_i \leq \delta(i,|H|)$$
$$\text{Subset } h: \qquad \delta(i,h) \leq \lambda_i + z_i - T_i \leq \delta(i,h-1)$$
$$\text{Subset } |H| + 1: \delta(i,1) \leq \lambda_i + z_i - T_i.$$

Besides the direct effects of coverage time on costs, the influence on downstream coverage times must be taken into account. Depending on the relation of z_j to the critical values, coverage times on a path to the final stage can always be larger than the times implying zero safety stock if z_j is not smaller than the corresponding critical value. Therefore, this case does not influence any concavity property on this path. If the relation $T_v > \tau_v$ is not feasible for a downstream stockpoint v, the coverage times on the path to v are set to τ and for v Case 1 applies. All decisions concerning stockpoints succeeding v are independent of the decision in i.

In order to allow for $T_i \geq \tau_i$, the critical values for z_i result from $\tau_i = \lambda_i + z_i - \delta(j)$. Under $z_i \geq 0$, $\delta(i) = \max\{0 \; ; \; \tau_i - \lambda_i + \delta(j)\} \; \forall \delta(j) \in H$ holds. Evaluating this recursive definition of the critical values gives the representation $\Delta(i,e)$ as used in Section 4.2.4.1. $\qquad \Box$

B. Minimal Coverage Properties

B.1 Serial Systems

Proposition B.1.1. *If a stockpoint $1 \leq r \leq n$ with*

$$\sum_{i=1}^{r} (\tau_i - \lambda_i) \geq 0 \qquad \text{and}$$

$$\sum_{i=1}^{q} (\tau_i - \lambda_i) < 0 \qquad \forall q < r$$

exists, stockpoints $1,...,r$ need no safety stock $(SS_i = 0 \ \forall i = 1,...,r)$ and can be excluded from further analysis.

Proof A feasible policy with zero costs for stockpoints $i = 1,...,r$ is given by

$$T_i = \min \left\{ \tau_i \ ; \ \lambda_i + \sum_{j=1}^{i-1} (\lambda_j - T_j) \right\} \qquad i = 1,...,r$$

because $T_i \leq \tau_i$ and therefore, $k_i = 0$ and $T_i \leq \lambda_i + \sum_{j=1}^{i-1} (\lambda_j - T_j)$. Additionally, the $\leq$ constraint belonging to stockpoint r in the optimization problem holds as an equality if the second condition from above is fulfilled. Every other policy causes higher costs at downstream stockpoints because the cost function is monotonously increasing in the coverage times. □

Proposition B.1.2. *For a non-reducible serial problem, the coverage times of the optimal policy satisfy*

$$T_i \geq \tau_i \qquad i = 1,...,n.$$

Proof Let j denote the most upstream stockpoint with $T_j < \tau_j$. T_j can be increased to τ_j without additional costs. Because Proposition B.1.1 is not applicable, there exist (one or more) stockpoints $i < j$ with $T_i > \tau_i$. The coverage times of these stockpoints can be decreased by the amount of $\tau_j - T_j$. This policy is always feasible and has lower costs than the original one with $T_j < \tau_j$. □

Theorem B.1.1. *Solving a non-reducible serial problem under γ-service level constraints, the optimal coverage times T_i^* obtain values resulting from the extreme points of the original solution set with additional constraints $T_i \geq \tau_i$.*

$$T_n^* = \lambda_n + \sum_{j=1}^{n-1}(\lambda_j - T_j^*) \qquad and$$

$$T_i^* \in \left\{ \tau_i \; ; \; \lambda_i + \sum_{j=1}^{i-1}(\lambda_j - T_j^*) - \Delta(i) \right\} \qquad i = 1,...,n-1$$

Proof If $\tau_i \leq \lambda_i \; \forall i = 1,...,n$, the constraints $T_i \geq \tau_i$ do not affect $\leq$ constraints in the original optimization problem. The only difference to the α-case is given by the modified lower values $T_i \geq \tau_i$. In the case $\tau_i > \lambda_i$ for at least one stockpoint i, all contraints of this type belonging to downstream stockpoints must be satisfied, which is only possible if $\Delta(i)$ remains uncovered after the evaluation of stockpoint i.   □

B.2 Divergent Systems

Proposition B.2.1. *If a stockpoint r with*

$$\sum_{j \in V(r)} (\tau_j - \lambda_j) \geq 0 \qquad and$$

$$\sum_{j \in V(q)} (\tau_j - \lambda_j) < 0 \qquad \forall q \in V(r)\backslash\{r\}$$

exists, then in the optimal policy, $SS_r = 0$. If $r \notin E$, the problem can be decomposed into $|n(r)| + 1$ subproblems of divergent structure where every direct successor of r is the root of a new divergent system and the original system is optimized without all stockpoints $\in N(r)$.

Proof To prove the proposition, we show that for an optimal policy, the safety stock of stockpoint r never covers more than τ_r periods and that the $\leq$ constraint belonging to stockpoint r in the divergent optimization problem is active.

First assume $T_r > \tau_r$. A policy with lower costs can be constructed by reducing T_r to τ_r and raising up predecessors of r with $T_j < \tau_j$ by the same amount which is possible since $\sum_{j \in V(r)}(\tau_j - \lambda_j) \geq 0$. Because of the second condition in Proposition B.2.1, no $\leq$ constraint belonging to preceding stockpoints is exceeded by this new policy. To satisfy all other constraints, some coverage times of successors to stockpoints with increased coverage times must be reduced. A policy constructed according to this logic is feasible and always has lower costs, since all reductions save costs and all increases result in no additional costs.

If the $\leq$ constraint belonging to stockpoint r in the divergent optimization problem is not active, T_r or coverage times of upstream stockpoints can be increased at no costs until the constraint is active. Simultaneously, costs are saved by reducing coverage times of succeeding stockpoints.   □

Proposition B.2.2. *For a non-decomposable divergent problem under γ-service level constraints, the optimal coverage times satisfy*

$$T_i^* \geq \min \left\{ \tau_i \; ; \; \lambda_i + \sum_{j \in V(i)\backslash\{i\}} (\lambda_j - T_j^*) \right\} \qquad i = 1, ..., n.$$

Proof Let i denote the most upstream stockpoint that does not satisfy the condition. Then, T_i can be increased to the minimum without any cost since $T_i \leq \tau_i$. This does not affect the respective $\leq$ constraint for i in the optimization problem because $T_i \leq \lambda_i + \sum_{j \in V(i)\backslash\{i\}} (\lambda_j - T_j)$. Since decomposition is not possible, a downstream coverage time reduction with a cost saving is possible because at least one successor stockpoint with $T_j > \tau_j$ exists.   □

B.3 Convergent Systems

Proposition B.3.1. *In the α-service level case, the maximum predecessor service time results from the path with the largest cumulative processing time.*

$$a(i) = \arg \max_{a \in A(i)} \left\{ \sum_{j \in w(a,i)} \lambda_j \right\}.$$

Proof Assume, that for a stockpoint i the maximum sum of uncovered processing times does not result from the path with the largest cumulative processing time (starting with $a(i)$). Because all processing times have to be covered, the maximum sum of uncovered processing times is covered by safety stocks at i or downstream stockpoints. On the path from $a(i)$ to i there is at least one coverage time $T_j > 0$ which can be reduced. Thereby, costs are saved because the excess coverage of stockpoint i or downstream stockpoints is large enough to enable this reduction.   □

Proposition B.3.2. *If there exist two stockpoints r and $a \in A(r)$ that satisfy*

$$\sum_{j \in w(a,r)} (\tau_j - \lambda_j) \geq 0 \text{ and}$$

$$\sum_{j \in w(a,q)} (\tau_j - \lambda_j) < 0 \; \forall q \in w(a,r)\backslash\{r\},$$

then stockpoint $a \in A$ needs no safety stock and is excluded from further analysis.

Proof To prove the proposition it is shown that if both conditions are satisfied, $T_a \leq \tau_a$ will always hold. The case $\tau_a \geq \lambda_a$ is trivial, since no safety stocks are necessary. If $T_a > \tau_a$, a reduction of T_a to τ_a saves costs and, because of the first condition, there exist (one or more) stockpoints $i \in w(a,r)$ with $T_i < \tau_i$. These T_i can be increased to τ_i at no cost. If a $\leq$-constraint in the original optimization problem is exceeded due to this, coverage times of other preceding stockpoints can be reduced, which again results in a cost reduction.   □

Proposition B.3.3. *In the γ-service level case, the maximum predecessor service time results from the path with the largest sum of processing times minus the times implying zero safety stocks.*

$$a(i) = \arg \max_{a \in A(i)} \{ \sum_{j \in w(a,i)} (\lambda_j - \tau_j) \}.$$

Proof Define $b(i) := \arg \max_{b \in A(i)} \{ \sum_{j \in w(b,i)} (\lambda_j - T_j) \}$. The sum of uncovered processing times on the path from $b(i)$ to i is covered by safety stocks held at i or successors to i. If $a(i) \neq b(i)$, there exists a stockpoint j on the path from $b(i)$ to i with $T_j < \tau_j$. A policy with lower costs can be constructed by increasing T_j to τ_j and reducing coverage times at i or successors to i. □

Proposition B.3.4. *For a non-reducible convergent problem, the coverage times of the optimal policy satisfy*

$$T_i \geq \tau_i \qquad\qquad i = 1, ..., n.$$

Proof Assume $T_i < \tau_i$. If a $\leq$ constraint in the convergent optimization problem is exceeded by an increase of T_i to τ_i, there exists one or more stockpoints with $T_j > \tau_j$ that can be decreased until all constraints are satisfied. Otherwise, a further problem reduction following Proposition B.3.2 is possible.   □

C. Numerical Algorithms

In cases where desired order-up-to-levels or safety stock parameters cannot be derived from a closed expression, numerical algorithms are necessary. In this appendix, two alternative methods are presented. The general problem being addressed is to find some value S that satisfies the equation $M(S) = x$ at an accuracy level of ϵ.

Bisection Procedure
    step 1: Find lower bound a and upper bound b
        $a := 0; b := \lambda \cdot \mu$
    while $M(b) < x$ **do**
        $a := b; b := b + \sigma\sqrt{\lambda}$
    step 2: Iteration $i = 1, 2, \ldots$
        repeat
        $S_i := \dfrac{a + b}{2}$
        if $M(S_i) < x$
            then $a := S_i$
            else $b := S_i$;
        until $|M(S_i) - x| < \epsilon$;

Newton-Raphson Method
    step 1: Compute starting value S_0
    step 2: Iteration $i = 1, 2, \ldots$
        while $|M(S_{i-1}) - x| > \epsilon$ **do**
        $S_i = S_{i-1} - (M(S_{i-1}) - x)/\frac{dM(S_{i-1})}{dS_{i-1}}$

This method will provide fast convergence. Covergence is guaranteed for normally distributed demands.[1] But, in general, this only holds if the starting value is sufficiently close to the desired solution.

---

[1] TIJMS, H.C., GROENEVELT, H. [1984], p. 181.

List of Symbols

Single-Echelon System

Demand

$D(t)$	demand random variable during t time periods
d_t	demand realization in period t
μ	single-period demand expectation
σ	single-period demand standard deviation
$g(i)$	autocovariance at lag i
v	coefficient of variation
$\hat{\mu}$	single-period demand expectation estimator
$\hat{\sigma}$	single-period demand standard deviation estimator
$\hat{g}(i)$	autocovariance at lag i estimator
$f(x)$	single-period demand probability density function
$F(x)$	single-period demand cumulative density function
$\varphi(x)$	single-period normal demand density
$\Psi(x)$	single-period normal demand cumulative density
$\phi(x)$	standard normal density
$\Phi(x)$	standard normal cumulative density
κ, q, l	parameters of Mixed-Erlang distribution
$e_{\kappa,l}(x)$	Erlang distribution density
$E_{\kappa,l}(x)$	Erlang distribution cumulative density
$ME_{\kappa,q,l}(x)$	Mixed-Erlang cumulative density

Lead Time

$\tilde{\lambda}$	processing time random variable
λ	processing time expectation
σ_λ	processing time standard deviation
f_λ	λ-period demand probability density
F_λ	λ-period demand cumulative probability density
$\hat{\lambda}$	planned processing time

Material Requirements

$a_{j,i}$ number of items of j required for one item of i

$\tilde{r}$ requirements level random variable for one item product outcome

r requirements level expectation

σ_r requirements level standard deviation

r_p planned requirements level

Inventory State and Replenishment Decisions

y_t net stock at the end of period t

x_t order quantity at the beginning of period t

OH_t on-hand stock at the end of period t

BL_t backlog at the end of period t

SH_t shortage quantity during period t

IP_t inventory position at the beginning of period t

s, S inventory control rule parameters

Service and Cost

K set-up cost

c variable production or purchasing cost per unit

p stockout penalty cost per unit and period

h inventory holding cost per unit and period

α, β, γ theoretical service levels

$\hat{\alpha}, \hat{\beta}, \hat{\gamma}$ realized empirical service levels

Safety Stock Determination

SS safety stock level

k safety factor

τ time covered by a zero safety stock level

$G(k)$ standardized loss function

$J(k)$ standardized quadratic loss function

w finite production rate

$\overline{Q}$ average lot-size quantity

LS_λ lost sales quantity over λ periods

Multi-Echelon System

The notation already introduced for the single-echelon system also applies
for the multi-echelon system with an additional stockpoint index.

Network

A	set of first-stage (most upstream) stockpoints without predecessors
E	set of final-stage (most downstream) stockpoints without successors
P	set of intermediate-stage stockpoints
$n(i)$	set of direct successors to stockpoint i
$v(i)$	set of direct predecessors to stockpoint i
$A(i)$	set of first-stage products that are required by i (identical to $\{i\}$ for $i \in A$)
$E(i)$	set of final-stage products that require product i (identical to $\{i\}$ for $i \in E$)
$N(i)$	set of all downstream stockpoints to i (including i)
$V(i)$	set of all upstream stockpoints to i (including i)
w	set of stockpoints, represents a path from a stockpoint to another downstream one
$w(i,j)$	set of stockpoints included in one path from i to j (including i and j)
$W(A,i)$	set of all paths that start at the first-stage and terminate in i
$W(A,E)$	set of all paths that start at the first-stage and terminate at the final stage
$a_{i,j}$	production coefficient, i.e. required material of product i in order to produce one unit of j

Demand

$\sigma_{i,j}$	covariance between single period demands of products i, j
$\rho_{i,j}$	demand correlation between single period demands of products i, j
$\hat{\rho}_{i,j}$	demand correlation estimator

Clark, Scarf Model

$\hat{y}_{i,t}$	echelon stock of i at the end of period t
$\tilde{y}_{i,t}$	echelon stock of i at the beginning of period t
$\hat{y}_{i,t}^{p}$	echelon inventory position of i at the end of t
$x_{i,t}$	order quantity of i at the beginning of t
$q_{i,t}$	shipment quantity of $i-1$ to i at the beginning of t

S_i	echelon order-up-to-level
Δ_i	maximum stock level at i
Y_i	shortfall random variable at i
$\tilde{F}^{[S_1,...,S_i]}$	shortfall distribution function at i
$F^{[S_1,...,S_i]}$	echelon stock distribution function at i
p_i	allocation fraction for requirements of i

Simpson Model

$\hat{y}_{i,t}$	projected net stock for i at the end of period t
$\hat{q}_{i,t}$	projected shipment to i at the beginning of t
$b_{i,t}$	stock insufficiency at i to fill the requirements of $i+1$
T_i	coverage time decision variable of i
ST_i	service time decision variable of i
L_i	replenishment lead time of i
B_i	base-stock level of i

Dynamic Programming Algorithm

$u_i,\underline{u}_i$	decision variable of i (single valued, vector valued)
$z_i,\underline{z}_i$	state variable of i (single valued, vector valued)
Λ	number of decision stages
Z_i	state space of i
$U_i(z_i),U_i(\underline{z}_i)$	decision space of i within a state z_i, $\underline{z}_i$, respectively
$f_i(z_i),f_i(\underline{z}_i)$	functional value of i in state z_i, $\underline{z}_i$, respectively

Local Search Procedures

$s(i)$	next safety stock holding successor to i
$p(i)$	next safety stock holding predecessor to i
$b(i)$	safety stock indicator for i
$\underline{b}$	solution representation for safety stock allocation
$NB(\underline{b})$	neighborhood to a solution $\underline{b}$
TP	annealing temperature
TH	threshold value
ΔC	cost difference between two solutions
PL	plateau length of evaluated solutions at a given temperature or threshold level
TL	Tabu list

Performance Measures

C_i^* cost value of the optimal solution of problem instance i

C_i^m cost value of the solution of problem instance i determined by method m

N number of problem instances

$\#opt^m$ number of optimal solutions found by method m

Δ_{max}^m maximum cost deviation of solutions obtained by method m

$\overline{\Delta}^m$ average cost deviation of solutions obtained by method m

CPU_{max} maximum computation time

$\overline{CPU}$ average computation time

Returns and Cyclic Network

F^E set of stockpoints that receive external returns

F^I set of stockpoints that receive internal returns

$R_{i,t}$ return random variable for stockpoint i in period t

$\mu_{i,bp}$ external return expectation at i

$\sigma_{i,bp}$ external return standard deviation at i

$bp(i)$ stockpoint of origin of returns that are reused by i

$a_{j,i}^{bp}$ by-product fraction of desired products at j (reused at i)

λ_i^{bp} reuse processing time for material reused at i

L_i^{bp} reuse mode replenishment lead time

$Q_{i,t}$ regular mode replenishment quantity of i at the beginning of t

List of Abbreviations

AR	Autoregressive process
c.f.	confer
CRP	Continuous Replenishment Process
DRP	Distribution Requirements Planning
E	Expectation
ECR	Efficient Consumer Response
EDI	Electronic Data Interchange
e.g.	for example
etc	etcetera
EOQ	Economic Order Quantity
EPQ	Economic Production Quantity
EQS	Economic Order Quantity under backordering
i.e.	id est
MA	Moving Average process
MHz	Mega-Hertz
MRP	Material Requirements Planning
p.	page
PC	Personal Computer
POS	Point-of-Sale
Prob	Probability
SIC	Stochastic Inventory Control
Var	Variance
VMI	Vendor Managed Inventories

List of Figures

List of Tables

References

[1] ABRAMOWITZ, M., STEGUN, I.A. [1970]: *Handbook of Mathematical Functions*. Dover Publications, New York

[2] ALSCHER, J., SCHNEIDER, H. [1982]: Zur Interdependenz von Fehlmengenkosten und Servicegrad. *Kostenrechnungspraxis*, 257-271

[3] AXSÄTER, S. [1993]: Continuous Review Policies for Multi-Level Inventory Systems with Stochastic Demand. In: GRAVES, S.C., RINNOOY KAN, A.H.G., ZIPKIN, P.H. (Eds.): *Logistics of Production and Inventory*. North-Holland, Amsterdam, 175-197

[4] AXSÄTER, S. [1997]: On Deficiencies of Common Ordering Policies for Multi-level Inventory Control. *OR Spektrum* 19, 109-110

[5] AXSÄTER, S., JUNTTI, L. [1996]: Comparison of Echelon Stock and Installation Stock Policies for Two-Level Inventory Systems. *International Journal of Production Economics* 45, 303-310

[6] AXSÄTER, S., JUNTTI, L. [1997]: Comparison of Echelon Stock and Installation Stock Policies with Policy Adjusted Order Quantities. *International Journal of Production Economics* 48, 1-6

[7] AXSÄTER, S., ROSLING, K. [1993]: Installation vs. Echelon Stock Policies for Multi-Level Inventory Control. *Management Science* 39, 1274-1280

[8] AXSÄTER, S., ROSLING, K. [1994]: Multi-Level Production-Inventory Control: Material Requirements Planning or Reorder Point Policies? *European Journal of Operational Research* 75, 405-412

[9] BADINELLI, R.D. [1996]: Approximating Probability Density Functions and Their Convolutions Using Orthogonal Polynomials. *European Journal of Operational Research* 95, 211-230

[10] BAKER, K.R. [1993]: Requirements Planning. In: GRAVES, S.C., RINNOOY KAN, A.H.G., ZIPKIN, P.H. (Eds.): *Logistics of Production and Inventory*. North-Holland, Amsterdam, 571-627

[11] BELLMAN, R. [1957]: *Dynamic Programming*. Princeton University Press, Princeton

[12] BELLMAN, R., GLICKSBERG, I., GROSS, O. [1955]: On the Optimal Inventory Equation. *Management Science* 2, 83-104

[13] BURGIN, T.A. [1975]: The Gamma Distribution and Inventory Control. *Operational Research Quarterly* 26, 507-525

206   References

[14] BUZACOTT, J.A., SHANTHIKUMAR, J.G. [1994]: Safety Stock Versus Safety Time in MRP Controlled Production Systems. *Management Science* 40, 1678-1689

[15] CHAKRAVARTY, A.K., SHTUB, A. [1986]: Simulated Safety Stock Allocation in a Two-Echelon Distribution System. *International Journal of Production Research* 24, 1245-1253

[16] CHARNES, J.M., MARMORSTEIN, H., ZINN, W. [1995]: Safety Stock Determination with Serially Correlated Demand in a Periodic-Review Inventory System. *Journal of the Operational Research Society* 46, 1006-1013

[17] CHIKAN, A. [1990]: *Inventory Models.* Kluwer, Dordrecht

[18] CLARK, A.J., SCARF, H. [1960]: Optimal Policies for a Multi-Echelon Inventory Problem. *Management Science* 6, 475-490

[19] CLARK, A.J., SCARF, H. [1962]: Approximate Solutions to a Simple Multi-Echelon Inventory Problem. In: Arrow, K.J., Karlin, S., Scarf, H. (Eds.): *Studies in Applied Probability and Management Science.* Stanford University Press, Stanford, 88-110

[20] DAS, C. [1978]: An Improved Formula for Inventory Decisions Under Service and Safety Stock Constraints. *AIIE Transactions* 10, 217-219

[21] DIKS, E.B. [1997]: *Controlling Divergent Multi-Echelon Systems*, PhD-Thesis Eindhoven University of Technology, Eindhoven

[22] DIKS, E.B., HEIJDEN, M.C. VAN DER [1997]: Modeling Stochastic Lead Times in Multi-Echelon Systems. *Probability in the Engineering and Informational Sciences* 11, 469-485

[23] DIKS, E.B., KOK, A.G. DE [1996]: Controlling a Divergent 2-Echelon Network with Transshipments Using the Consistent Appropriate Share Rationing Policy. *International Journal of Production Economics* 45, 369-379

[24] DIKS, E.B., KOK, A.G. DE [1998]: Transshipments in a Divergent 2-Echelon System. In: FLEISCHMANN, B., NUNEN, J.A.E.E. VAN, SPERANZA, M.G., STÄHLY, P. (Eds.): *Advances in Distribution Logistics.* Springer, Berlin Heidelberg New York, 423-447

[25] DIKS, E.B., KOK, A.G. DE, LAGODIMOS, A.G. [1996]: Multi-Echelon Systems: A Service Measure Perspective. *European Journal of Operational Research* 95, 241-263

[26] DONSELAAR, K. VAN [1989]: *Material Coordination Under Uncertainty.* PhD-Thesis, Eindhoven University of Technology, Eindhoven

[27] DONSELAAR, K. VAN [1990]: Integral Stock Norms in Divergent Systems with Lot-Sizes. *European Journal of Operational Research* 45, 70-84

[28] DUECK, G. [1993]: New Optimization Heuristics. The Great Deluge Algorithm and the Record-to-Record Travel. *Journal of Computational Physics* 104, 86-92

[29] DUECK, G., SCHEUER, T. [1990]: Threshold Accepting: A General Purpose Optimization Algorithm Appearing Superior to Simulated Annealing. *Journal of Computational Physics* 90, 161-175

[30] EGLESE, R.W. [1990]: Simulated Annealing: A Tool for Operational Research. *European Journal of Operational Research* 46, 271-281

[31] EL-NAJDAWI, M.K. [1993]: Empirical Evaluation of Safety Stock Policies. *International Journal of Management and Systems* 9, 277-286

[32] EPPEN, G.D. [1979]: Effects of Centralization on Expected Costs in a Multi-Location Newsboy Problem. *Management Science* 25, 498-501

[33] EPPEN, G.D., MARTIN, R.K. [1988]: Determining Safety Stock in the Presence of Stochastic Lead Time and Demand. *Management Science* 34, 1380-1390

[34] EPPEN, G., SCHRAGE, L. [1981]: Centralized Ordering Policies in a Multiwarehouse System with Leadtimes and Random Demand. In: SCHWARZ, L. (Ed.): *Multi-Level Production/Inventory Control Systems: Theory and Practice.* North-Holland, Amsterdam, 51-69

[35] ETIENNE, E.C. [1987]: Choosing Optimal Buffering Strategies for Dealing with Uncertainty in MRP. *Journal of Operations Management* 7, 107-120

[36] EVERS, P.T. [1995]: Expanding the Square Root Law: An Analysis of Both Safety and Cycle Stocks. *The Logistics and Transportation Review* 31, 1-20

[37] EVERS, P.T., BEIER, F.J. [1993]: The Portfolio Effect and Multiple Consolidation Points: A Critical Assessment of the Square Root Law. *Journal of Business Logistics* 14, 109-125

[38] EYNAN, A. [1996]: The Impact of Demands' Correlation on the Effectiveness of Component Commonality. *International Journal of Production Research* 34, 1581-1602

[39] FEDERGRUEN, A. [1993]: Centralized Planning Models for Multi-Echelon Inventory Systems Under Uncertainty. In: GRAVES, S.C., RINNOOY KAN, A.H.G., ZIPKIN, P.H. (Eds.): *Logistics of Production and Inventory.* North-Holland, Amsterdam, 133-174

[40] FEDERGRUEN, A., ZIPKIN, P. [1984a]: Computational Issues in an Infinite-Horizon, Multiechelon Inventory Model. *Operations Research* 32, 818-836

[41] FEDERGRUEN, A., ZIPKIN, P. [1984b]: Approximations of Dynamic, Multilocation Production and Inventory Problems. *Management Science* 30, 69-84

[42] FEDERGRUEN, A., ZIPKIN, P. [1984c]: Allocation Policies and Cost Approximations for Multilocation Inventory Systems. *Naval Research Logistics Quarterly* 31, 97-129

[43] FISHER, M.L. [1997]: What is the Right Supply Chain for Your Product? *Harvard Business Review* March-April, 105-116

[44] FISHER, M., RAMAN, A. [1996]: Reducing the Cost of Demand Uncertainty Through Accurate Response to Early Sales. *Operations Research* 44, 87-99

[45] FORRESTER, J.W. [1961]: *Industrial Dynamics*. MIT Press, Cambridge

[46] GLASSERMAN, P. [1997]: Bounds and Asymptotics for Planning Critical Safety Stocks. *Operations Research* 45, 244-257

[47] GLOVER, F. [1989]: Tabu Search-Part I. *ORSA Journal on Computing* 1, 190-206

[48] GLOVER, F. [1990]: Tabu Search-Part II. *ORSA Journal on Computing* 2, 4-32

[49] GLOVER, F. [1993]: A User's Guide to Tabu Search. *Annals of Operations Research* 41, 3-28

[50] GRAVES, S.C. [1988]: Safety Stocks in Manufacturing Systems. *Journal of Manufacturing and Operations Management* 1, 67-101

[51] GRAVES, S.C., WILLEMS S.P. [1998]: Optimizing Strategic Safety Stock Placement in Supply Chains. Working Paper A.P. Sloan School of Management, Massachusetts Institute of Technology

[52] GÜLLÜ, R. [1996]: On the Value of Information in Dynamic Production/Inventory Problems under Forecast Evolution. *Naval Research Logistics* 43, 289-303

[53] GÜLLÜ, R. [1997]: A Two-Echelon Allocation Model and the Value of Information Under Correlated Forecasts and Demands. *European Journal of Operational Research* 99, 386-400

[54] HADLEY, G., WHITIN, T.M. [1963]: *Analysis of Inventory Systems*. Prentice Hall, Englewood Cliffs

[55] HASTINGS, C. [1955]: *Approximations for Digital Computers*. Princeton, New Jersey

[56] HAX, A.C., CANDEA, D. [1984]: *Production and Inventory Management*. Prentice Hall, Englewood Cliffs

[57] HEATH, C.D., JACKSON, P.L. [1991]: Modelling the Evolution of Demand Forecasts with Application to Safety Stock Analysis in Production/Distribution Systems. *IIE Transactions* 26, 17-30

[58] HEIJDEN, M.C. VAN DER [1997a]: Supply Rationing in Multi-Echelon Divergent Systems. *European Journal of Operational Research* 101, 532-549

[59] HEIJDEN, M.C. VAN DER [1997b]: Multi-Echelon Inventory Control in Divergent Systems with Shipping Frequencies. Working Paper P5, Institute for Business Engineering and Technology Application, University of Twente

[60] HEIJDEN, M.C. VAN DER, DIKS, E.B., KOK, A.G. DE [1997]: Stock Allocation in General Multi-Echelon Distribution Systems with (R,S) Order-up-to-Policies. *International Journal of Production Economics* 49, 157-174

[61] HEIJDEN, M.C. VAN DER, DIKS, E.B., KOK, A.G. DE [1999]: Inventory Control in Multi-Echelon Divergent Systems with Random Lead Times. *OR Spektrum* 21, 331-359

[62] HEINRICH, C.E. [1987]: *Mehrstufige Losgrößenplanung in hierarchisch strukturierten Produktionsplanungssystemen.* Springer, Berlin Heidelberg New York

[63] HEINRICH, C.E., SCHNEEWEISS, C. [1986]: Multi-Stage Lot-Sizing for General Production Systems. In: AXSÄTER, S., SCHNEEWEISS, C., SILVER, E.A. (Eds.): *Multi-Stage Production Planning and Inventory Control.* Springer, Berlin Heidelberg New York, 150-181

[64] HORST, R., TUY, H. [1996]: *Global Optimization.* 3rd ed., Springer, Berlin Heidelberg New York

[65] HOUTUM, G.J. VAN, INDERFURTH, K., ZIJM, W.H.M. [1996]: Materials Coordination in Stochastic Multi-Echelon Systems. *European Journal of Operational Research* 95, 1-23

[66] HOUTUM, G.J. VAN, ZIJM, W.H.M. [1991]: Computational Procedures for Stochastic Multi-Echelon Production Systems. *International Journal of Production Economics* 23, 223-237

[67] HOUTUM, G.J. VAN, ZIJM, W.H.M. [1997]: Incomplete Convolutions in Production and Inventory Models. *OR Spektrum* 19, 97-107

[68] IGLEHART, D.L. [1963]: Optimality of (S, s) Policies in the Infinite Horizon Dynamic Inventory Problem. *Manangement Science* 9, 259-267

[69] INDERFURTH, K. [1991]: Safety Stock Optimization in Multi-Stage Inventory Systems. *International Journal of Production Economics* 24, 103-113

[70] INDERFURTH, K. [1992]: Mehrstufige Sicherheitsbestandsplanung mit dynamischer Programmierung. *OR Spektrum* 14, 19-32

[71] INDERFURTH, K. [1993]: Valuation of Leadtime Reduction in Multi-Stage Production Systems. In: FANDEL, G., GULLEDGE, T., JONES, A. (Eds.): *Operations Research in Production Planning and Inventory Control.* Springer, Berlin Heidelberg New York, 413-427

[72] INDERFURTH, K. [1994]: Safety Stocks in Multi-Stage Divergent Inventory Systems: A Survey. *International Journal of Production Economics* 35, 321-329

[73] INDERFURTH, K. [1995]: Multistage Safety Stock Planning with Item Demands Correlated Across Products and Through Time. *Production and Operations Management* 4, 127-144

[74] IYER, A.V., BERGEN, M.E. [1997]: Quick Response in Manufacturer-Retailer Channels. *Management Science* 43, 559-570

[75] JANSSEN, F.B.S.L.P. [1998]: *Inventory Management Systems.* PhD-Thesis, Tilburg University, Tilburg

[76] JENSEN, T. [1996]: *Planungsstabilität in der Material-Logistik.* Physica, Heidelberg

[77] KÄSSMANN, G., KÜHN, M., SCHNEEWEISS, CH. [1986]: Spicher's SB-Algorithmus Revisited - Feedback versus Feedforeward - Steuerung in der Lagerhaltung. *OR Spektrum* 8, 89-98

[78] KARLIN, S. [1958]: Steady State Solutions. In: ARROW, K.A., KARLIN, S., SCARF, H. (Eds.): *Studies in the Mathematical Theory of Inventory and Production.* Stanford University Press, Stanford, Chapter 14

[79] KARMARKAR, U.S. [1987]: Lot Sizes, Lead Times and In-Process Inventories. *Management Science* 33, 409-418

[80] KARMARKAR, U.S. [1993]: Manufacturing Lead Times, Order Release and Capacity Loading. In: GRAVES, S.C., RINNOOY KAN, A.H.G., ZIPKIN, P.H. (Eds.): *Logistics of Production and Inventory.* North-Holland, Amsterdam, 287-329

[81] KELLE, P., SILVER, E.A. [1990]: Safety Stock Reduction by Order Splitting. *Naval Research Logistics* 37, 725-743

[82] KIMBALL, G.E. [1988]: General Principles of Inventory Control. *Journal of Manufacturing and Operations Management* 1, 119-130

[83] KLEIJN, M.J., DEKKER, R. [1998]: Using Break Quantities for Tactical Optimisation in Multi-Stage Distribution Systems. In: FLEISCHMANN, B., NUNEN, J.A.E.E. VAN, SPERANZA, M.G., STÄHLY, P. (Eds.): *Advances in Distribution Logistics.* Springer, Berlin Heidelberg New York, 305-317

[84] KOK, A.G. DE [1990]: Hierarchical Production Planning for Consumer Goods. *European Journal of Operational Research* 45, 55-69

[85] KOK, A.G. DE, LAGODIMOS, A.G., SEIDEL, H.P. [1994]: Stock Allocation in a 2-Echelon Distribution Network Under Service Contraints. Research Report TUE/BDK/LBS 94-03, Eindhoven University of Technology

[86] KOK, A.G. DE, VISSCHERS, J.W.C.H. [1999]: Analysis of Assembly Systems with Service Level Constraints. *International Journal of Production Economics* 59, 313-326

[87] LAGODIMOS, A.G. [1990]: *Protective Inventories in Manufacturing Systems.* PhD-Thesis, University of Cambridge, Cambridge

[88] LAGODIMOS, A.G. [1992]: Multi-Echelon Service Models for Inventory Systems Under Different Rationing Policies. *International Journal of Production Research* 30, 939-958

[89] LAGODIMOS, A.G. [1993]: Models for Evaluating the Performance of Serial and Assembly MRP Systems. *European Journal of Operational Research* 68, 49-68

[90] LAGODIMOS, A.G., ANDERSON, E.J. [1993]: Optimal Positioning of Safety Stocks in MRP. *International Journal of Production Research* 31, 1797-1813

[91] LAGODIMOS, A.G., KOK, A.G. DE, VERRIJDT, J.H.C.M [1995]: The Robustness of Multi-Echelon Service Models Under Autocorrelated Demands. *Journal of the Operational Research Society* 46, 92-103

[92] LAMBRECHT, M.R., MUCKSTADT, J.A., LUYTEN, R. [1984]: Protective Stocks in Multi-Stage Production Systems. *International Journal of Production Research* 22, 1001-1025

[93] LANGENHOFF, L.J.G., ZIJM, W.H.M. [1990]: An Analytical Theory of Multi-Echelon Production/Distribution Systems. *Statistica Neerlandica* 44, 149-174

[94] LAU, H.-S., ZAKI, A. [1982]: The Sensitivity of Inventory Decisions to the Shape of Lead Time-Demand Distribution. *IIE Transactions* 14, 265-271

[95] LEE, H.L., BILLINGTON, C. [1992]: Managing Supply Chain Inventory: Pitfalls and Opportunities. *Sloan Management Review* Spring, 65-73

[96] LEE, H.L., BILLINGTON, C. [1993]: Material Management in Decentralized Supply Chains. *Operations Research* 41, 835-847

[97] LEE, H.L., NAHMIAS, S. [1993]: Single-Product, Single-Location Models. In: GRAVES, S.C., RINNOOY KAN, A.H.G., ZIPKIN, P.H. (Eds.): *Logistics of Production and Inventory.* North-Holland, Amsterdam, 3-55

[98] LEE, H.L., PADMANABHAN, V., WHANG, S. [1997]: The Bullwhip Effect in Supply Chains. *Sloan Management Review* Spring, 93-102

[99] LEE, H.L., TANG, C.S. [1997]: Modelling the Costs and Benefits of Delayed Product Differentiation. *Management Science* 43, 40-53

[100] LUYTEN, R. [1986]: System-Based Heuristics for Multi-Echelon Distribution Systems. In: AXSÄTER, S., SCHNEEWEISS, C., SILVER, E.A. (Eds.): *Multi-Stage Production Planning and Inventory Control.* Springer, Berlin Heidelberg New York, 50-91

[101] MAGEE, J.F., BOODMAN, D.M. [1967]: *Production Planning and Inventory Control.* 2nd ed., McGraw-Hill, New York

[102] MOLINDER, A. [1997]: Joint Optimization of Lot-Sizes, Safety Stocks and Safety Lead Times in an MRP System. *International Journal of Production Research* 35, 983-994

[103] MUCKSTADT, J.A., ROUNDY, R.O. [1993]: Analysis of Multistage Production Systems. In: GRAVES, S.C., RINNOOY KAN, A.H.G., ZIPKIN, P.H. (Eds.): *Logistics of Production and Inventory.* North-Holland, Amsterdam, 59-131

[104] NAHMIAS, S. [1979]: Simple Approximations for a Variety of Dynamic Leadtime Lost-Sales Inventory Models. *Operations Research* 27, 904-924

[105] NATARAJAN, R., GOYAL, S.K. [1994]: Safety Stocks in JIT Environments. *International Journal of Operations and Production Management* 14, No. 10, 64-71

[106] NEW, C. [1975]: Safety Stocks for Requirements Planning. *Production and Inventory Management* 16, 1-18

[107] ORLICKY, J. [1975]: *Material Requirements Planning.* McGraw-Hill, New York

[108] PIRLOT, M. [1996]: General Local Search Methods. *European Journal of Operational Research* 92, 493-511

[109] RAY, W.D. [1980]: The Significance of Correlated Demands and Variable Lead Times for Stock Control Policies. *Journal of the Operational Research Society* 31, 187-190

[110] RAY, W.D. [1981]: Computation of Reorder Levels when the Demands are Correlated and the Lead Time Random. *Journal of the Operational Research Society* 32, 27-34

[111] RITCHKEN, P.H., SANKAR, R. [1984]: The Effect of Estimation Risk in Establishing Safety Stock Levels in an Inventory Model. *Journal of the Operational Research Society* 35, 1091-1099

[112] ROSLING, K. [1989]: Optimal Inventory Policies for Assembly Systems Under Random Demands. *Operations Research* 37, 565-579

[113] SALOMON, M., KUIK, R., VAN WASSENHOVE, L.N. [1993]: Statistical Search Methods for Lotsizing Problems. *Annals of Operations Research* 41, 453-468

[114] SCARF, H. [1960]: The Optimality of (S, s) Policies in the Dynamic Inventory Problem. In: ARROW, K.J., KARLIN, S., SUPPES, P. (Eds.): *Mathematical Methods in the Social Sciences*. Stanford University Press, Stanford, 196-202

[115] SCHILDT, B. [1994]: *Strategische Produktions- und Distributionsplanung*. Deutscher Universitäts Verlag, Wiesbaden

[116] SCHMIDT, C.P, NAHMIAS, S. [1985]: Optimal Policy for a Two-Stage Assembly System Under Random Demand. *Operations Research* 33, 1130-1145

[117] SCHNEEWEISS, C. [1999]: *Hierarchies in Distributed Decision Making*. Springer, Berlin Heidelberg New York

[118] SCHNEIDER, H. [1978]: Methods for Determining the Re-Order Point of an (s, S) Ordering Policy when a Service Level is Specified. *Journal of the Operational Research Society* 29, 1181-1193

[119] SCHNEIDER, H. [1979]: *Servicegrade in Lagerhaltungsmodellen*. Marchal und Matzenbacher, Berlin

[120] SCHNEIDER, H. [1981]: Effect of Service-Levels on Order-Points or Order-Levels in Inventory Models. *International Journal of Production Research* 19, 615-631

[121] SCHNEIDER, H., RINKS, D.B., KELLE, P. [1995]: Power Approximations for a Two-Echelon Inventory System Using Service Levels. *Production and Operations Management* 4, 381-400

[122] SEIDEL, H.P., DE KOK, A.G. [1990]: Analysis of Stock Allocation in a 2-Echelon Distribution System. Technical Report 098, CQM, Eindhoven

[123] SEGERSTEDT, A. [1995]: Cover-Time Planning, a Method for Calculation of Material Requirements. *International Journal of Production Economics* 41, 355-368

[124] SHERBROOKE, C.C. [1992]: *Optimal Inventory Modeling of Systems*. Wiley, New York

[125] SILVER, E.A., PYKE, D.F., PETERSON, R. [1998]: *Inventory Management and Production Planning and Scheduling.* 3rd ed., Wiley, New York

[126] SIMPSON, K.F. [1958]: In-Process Inventories. *Operations Research 6,* 863-873

[127] SIMPSON, V.P. [1976]: A Noniterative Approximation to the Optimum *EOQ* with Service and Nonnegative Safety Stock Constraints. *AIIE Transactions 8,* 155-157

[128] SNYDER, R.D. [1980]: The Safety Stock Syndrome. *Journal of the Operational Research Society 31,* 833-837

[129] SOUTH, J.B., HIXSON, R. [1988]: Excess Capacity Versus Finished Goods Safety Stock. *Production and Inventory Management,* 3rd Quarter, 36-40

[130] SPICHER, K. [1975]: Der SB_1-Algorithmus - Eine Methode zur Beschreibung des Zusammenhangs zwischen Ziel-Lieferbereitschaft und Sicherheitsbestand. *Zeitschrift für Operations Research 19,* B1-B12

[131] SPICHER, K. [1976]: Erfahrungen mit dem SB_1-Algorithmus. *Zeitschrift für Operations Research 20,* B115-B126

[132] SUCHANEK, B. [1996]: *Sicherheitsbestände zur Einhaltung von Servicegraden.* Lang, Frankfurt am Main

[133] SYDSAETER, K., HAMMOND, P.J. [1995]: *Mathematics for Economic Analysis.* Prentice Hall, Englewood Cliffs

[134] TEMPELMEIER, H. [1985]: Inventory Control Using a Service Constraint on the Expected Customer Order Waiting Time. *European Journal of Operational Research 19,* 313-323

[135] TIJMS, H.C. [1994]: *Stochastic Models: An Algorithmic Approach.* Wiley, Chichester

[136] TIJMS, H.C., GROENEVELT, H. [1984]: Simple Approximations for the Reorder Point in Periodic and Continuous Review (s,S) Inventory Systems with Service Level Constraints. *European Journal of Operational Research 17,* 175-190

[137] TÜSHAUS, U., WAHL, C. [1998]: Inventory Positioning in a Two-Stage Distribution System with Service Level Constraints. In: FLEISCHMANN, B., NUNEN, J.A.E.E. VAN, SPERANZA, M.G, STÄHLY, P. (Eds.): *Advances in Distribution Logistics.* Springer, Berlin Heidelberg New York, 501-529

[138] VERRIJDT, J.H.C.M., KOK, A.G. DE [1995]: Distribution Planning for a Divergent N-Echelon Network Without Intermediate Stocks Under Service Restrictions. *International Journal of Production Economics 38,* 225-243

[139] VERRIJDT, J.H.C.M., KOK, A.G. DE [1996]: Distribution Planning for a Divergent Depotless Two-Echelon Network Under Service Constraints. *European Journal of Operational Research 89,* 341-354

[140] WHITTEMORE, A.S., SAUNDERS, S.C. [1977]: Optimal Inventory Under Stochastic Demand with Two Supply Options. *SIAM Journal of Applied Mathematics* 32, 293-305

[141] WHYBARK, D.C., WILLIAMS, J.G. [1976]: Material Requirements Planning Under Uncertainty. *Decision Sciences* 7, 595-606

[142] WIJNGAARD, J., WORTMANN, J.C. [1985]: MRP and Inventories. *European Journal of Operational Research* 20, 281-293

[143] YANO, C.A., LEE, H.L. [1995]: Lot Sizing with Random Yields: A Review. *Operations Research* 43, 311-334

[144] ZIJM, W.H.M., HOUTUM, G.J. VAN [1994]: On Multi-Stage Production/Inventory Systems Under Stochastic Demand. *International Journal of Production Economics* 35, 391-400

[145] ZIPKIN, P. [1984]: On the Imbalance of Inventories in Multi-Echelon Systems. *Mathematics of Operations Research* 9, 402-423

Printing: Weihert-Druck GmbH, Darmstadt
Binding: Buchbinderei Schäffer, Grünstadt